T0331622

Case Studies in Suburban Sustainability

CASE STUDIES IN SUBURBAN SUSTAINABILITY

EDITED BY

Sandra J. Garren and Robert Brinkmann

University of Florida Press

Gainesville

25 24 23 22 21 20   6 5 4 3 2 1

Library of Congress Cataloging-in-Publication Data
Names: Garren, Sandra J., editor. | Brinkmann, Robert, 1961– editor.
Title: Case studies in suburban sustainability / edited by Sandra J.
    Garren, Robert Brinkmann.
Description: Gainesville : University of Florida Press, 2020. | Includes
    bibliographical references and index.
Identifiers: LCCN 2020017102 (print) | LCCN 2020017103 (ebook) | ISBN
    9781683401599 (hardback) | ISBN 9781683401872 (pdf)
Subjects: LCSH: Sustainable development—United States—Case studies. |
    Sustainable urban development—United States—Case studies. | Community
    development, Urban—United States—Case studies.
Classification: LCC HC79.E5 C3738 2020 (print) | LCC HC79.E5 (ebook) |
    DDC 307.1/4140973—dc23
LC record available at https://lccn.loc.gov/2020017102
LC ebook record available at https://lccn.loc.gov/2020017103

University of Florida Press
2046 NE Waldo Road
Suite 2100
Gainesville, FL 32609
http://upress.ufl.edu

UF PRESS

UNIVERSITY
OF FLORIDA

CONTENTS

FIGURES

TABLES

CASE STUDIES IN SUBURBAN SUSTAINABILITY

1

Introduction

The State of Suburban Sustainability in the United States

ROBERT BRINKMANN AND SANDRA J. GARREN

The suburbs might seem like an odd place to focus a book on sustainability. They are often derided as unsustainable landscapes where people dispro-portionately use resources compared to their urban counterparts (Kahn 2000). They are rarely considered in the sustainability literature, where in terms of human settlements, the focus tends to be on cities or metropolitan places that bring together suburbs with their urban counterparts (Rose-land 2012). The suburbs are seldom examined as a whole. While suburbs are sometimes examined in a thematic context, such as homes, small farms, or unique transit opportunities, overall they receive little attention. This book tries to remedy this issue by addressing sustainability within a uniquely sub-urban context.

It is important to examine suburbs in this critical time. Suburbs are changing (Niedt 2013). When they first started to transform the American landscape in the 1940s and 1950s, many white Americans migrated to sub-urbs as part of "white flight" and to escape the complexities of urban life (Boustan 2010). Some suburbs became monochromatic bedroom commu-nities that redefined America. The suburban migration, or "white flight," left behind shattered and defunded cities that took decades to recover as wealthier residents left the city behind (Schneider 2008).

The structure of suburbs, even with recent attempts to enhance densifica-tion, is characterized by low density with the limited presence of undesir-able attributes such as power plants, waste dumps, or polluting industries (Short et al. 2007). Certainly, there are notable examples of suburbs that were constructed on polluted lands (Love Canal for example [Gibbs 2011]), but the overall characterization of suburbs is one that is green and clean.

The suburbs are also places with limited tax burdens (Brueckner 2000). They usually have lower tax rates compared to their urban counterparts. This means that governments and services are limited. There are usually few amenities, particularly publicly funded amenities. The suburbs were designed by and for people who do not need many services and who do not want many amenities.

But things are changing significantly. The migration trend is shifting. Many now prefer to live in cities to be closer to work and to have access to services, public transportation, and amenities (Hyra 2015). Young people in particular are drawn to the new dynamism in urban centers (Myers 2016). In many places in the United States, particularly the rust belt, older suburbs are in decline while in other places, especially in the Sunbelt, newer suburbs are thriving along with their urban counterparts. The suburbs are increasingly becoming far more dynamic places than they have been in the past (Strom 2017).

This dynamism is expressed in interesting ways. For example, the suburbs are becoming more diverse ethnically, socially, politically, and economically. Take for example the area around our academic home, Hofstra University. We are a stone's throw from Levittown, New York, and other suburban planned communities. These were home of some of America's first Fordist suburban developments where there were covenants to exclude particular ethnic groups such as African Americans and Jewish people. Today, however, the suburban landscape of Long Island is a patchwork of diversity. Hicksville, for example, is home to a thriving South Indian community, and many places across the island have welcomed new immigrants from Central America and Ecuador. Nationally, the suburbs are often the first choice of new immigrants (Walker 2014). We have moved from the time when new immigrants flocked to cities (the Lower East Side of New York [Foner 2000] for example or Little Havana in Miami [Duany 2011]). Today, they are choosing to move to suburban Hempstead, New York, or Homestead, Florida, on the edges of their larger urban counterpart. Unfortunately, these changes in the suburbs have resulted in enhanced segregation *within* the suburbs (Anacker et al. 2017).

This dynamism of diversity has kept suburbs populated even as they age. For example, suburban Los Angeles continues to grow, in part, due to immigration from Mexico and other Pacific Rim nations (Soja 2014). Yet these suburbs continue to have limited tax bases and thus limited support for services and amenities. Thus, as they age, there is limited ability to reinvest in new roads, sewers, schools, or social services. That is why some suburbs are left behind in decline as new suburbs are constructed. There is a constant demand for new housing as older suburbs fade.

The growth and recent dynamic changes taking place in suburbs provides a very fruitful ground for research. Many new important books have refocused attention on suburban landscapes. This book adds to the growing literature by looking at sustainability in the American suburb. Before delving into why the suburbs are emerging in importance for sustainability research, it is worth examining the development of sustainability as a discipline. It is a relatively new field that emerged in the late 1980s as new evidence for global human-induced environmental change came forward.

DEVELOPMENT AND GROWTH OF SUSTAINABILITY

The American environmental movement is complex. In part, it evolved out of mid-nineteenth century thought that focused on nature as something that could elevate human existence. This was partially a result of the description, drawings, paintings, and early photographs of dramatic landscapes (Novak 2007). The Romantic Movement drew upon the ideal of nature and created a certain landscape aesthetic that can be represented in the Hudson Valley School of Art and the western paintings of George Catlin. The transcendental writings of Thoreau can be seen as part of this continuum that placed nature as a way to create individual meaning and connection with one's own existence (Higgins 2017).

However, as late nineteenth- and early twentieth-century industrial development ventured west, the spectacular landscapes were under threat from development. There was a movement to preserve and protect lands through the development of national parks and forests. This new dynamic set up a new movement whereby nature was managed by experts and also created a divide between preservation (parks) and multiple-use (forests) advocates who sought different outcomes from public lands. This divide was fundamentally expressed through the Hetch Hetchy controversy around the development of a national park that pitted preservationist John Muir against multiple-use advocate Gifford Pinchot (Righter 2005). This tension remains today as we see federal administrations change policies with elections. The modification of the borders of the national monuments and the incursion of oil, gas, mining, and agricultural interests on federal lands show that this debate rages on into our present era.

By the 1950s, however, a new theme was finding its way into the environmental discourse. It was quite clear by that time that widespread industrialization was significantly altering the environment and harming not only ecosystems, but also human health. The work of Rachel Carson and Aldo Leopold exemplify the type of work that was taking place during this period. Leopold specifically focused on changing ecosystems (Leopold 1949) while Carson placed emphasis on how we were chemically altering the natural world through industrial chemicals (Carson 1962). Outcomes of this era included the development of laws and organizations that managed preservation and pollution. The Clean Water Act, the Endangered Species Act, and the Clean Air Act are just a few examples of the types of laws that were enacted at the federal level. As a result of these policies, the United States public saw great improvements in water and air quality and greater interest in preservation. These outcomes, while important, were not felt at the global level, and it was clear that there were global environmental threats that could not be addressed by national laws.

The first major global challenge that caught the attention of the scientific and policy community was the issue of chlorofluorocarbon (CFC) pollution and its impact on the ozonosphere (Murdoch and Sandler 1997). While ozone is a pollutant at lower levels, it protects the planet from harmful radiation in the upper layers of the atmosphere. In the 1970s and 1980s, it became evident that CFCs emitted in aerosols such as hairspray and deodorant were damaging this high ozonosphere. The only way to stop the problem was to develop a global approach to ban CFCs. If one country banned the chemical and other countries continued to use it, the problem would still exist. The world needed a global approach.

Thankfully one emerged in 1987, the Montreal Protocol, that required signatories to reduce the use and production of chemicals harmful to the ozone layer (United Nations Environment Programme 2018). The effort showed that the world could come together to address global environmental issues. As the world took stock of the ozone issue, other global environmental issues started to garner significant attention. Climate change, food production and distribution, water quantity and quality, income inequality, ecosystem degradation, ocean pollution, overfishing, and many other problems began to be seen not just as local problems but also as global issues that could be solved if the world could work together. It was evident that these environmental problems were accelerating as human population and economic activity increased in the later twentieth century. There was great concern as to whether or not we could survive as a species as we altered the planet in ways that were unimaginable a generation before. In short, we became concerned with the *sustainability* of the planet.

The idea of environmental sustainability was not new (the Tragedy of the Commons [Hardin 1968] provides a good primer), but it was repackaged within the global context as the United Nations started to take up global challenges of environmental sustainability in the 1980s at a time when much of the world was going through significant growth. In 1983, the United Nations formed a commission at the direction of its secretary general, Javier Perez de Cuellar de la Guerra, to address global environmental change within the context of development. The group was called the World Commission on Environment and Development and was chaired by former Norwegian prime minister Gro Harlem Brundtland.

The commission looked at a variety of global environmental problems and came to the conclusion that there were three main attributes that must be examined when considering the sustainability of the planet. These attributes, now called the three pillars of sustainability, focus not only on environmental issues, but also on social and economic considerations. This change of approach had a significant impact on how much of the world looked at environmental problems. Instead of seeing them as scientific challenges that could be engineered through laws and technology, they were seen more broadly within a cultural context (Brundtland Commission 1987). The report also made a strong case for considering future generations within its strong definition of sustainable development as meeting the needs of the present without limiting the ability of future generations to meet their needs.

Since the publication of the Brundtland Report in 1987 which first defined sustainability as used in our present context, the field of sustainability has been expanding tremendously, especially as we confront issues in the United States such as climate change, environmental justice, ecosystem decline, and increased pollution of materials like plastics and hormones that were not regulated in the initial rule making that started in the 1960s.

The United Nations has been extremely active in sustainability over the last several decades. For example, the Earth Summit in 1992 produced three important documents that provided guidance for future years: *Rio Declaration on Environment and Development, Agenda 21, and Forest Principles*. Perhaps the most significant of these in terms of the suburbs was *Agenda 21* which provided nonbinding guidance for sustainable development in communities. Since the Earth Summit, several other important events have been held, but perhaps the most significant development for the suburbs was the adoption of the Sustainable Development Goals (SDGs) by the United Nations in 2015. These 17 goals (Figure 1.1) cover a myriad of sustainability issues that are framed within measurable outcomes.

There are two goals, SDG 11 Sustainable Cities and Communities and SDG 17 Partnerships for the Goals, that are of particular interest to the topic of suburban sustainability (Figure 1.1).

These targets can easily be applied within a suburban context for framing sustainability initiatives. SDG 17, which focuses on building partnerships, contains far more targets that can be reviewed here, but they are divided into the following areas: finance (largely focused on financial sustainability and finance for sustainability), technology, capacity building, trade, and systemic issues which includes policy and institutional coherence, multistakeholder partnerships, and data monitoring and accountability. While many look at the SDGs as a tool for national-scale sustainability, the reality is that they were developed with communities in mind in order for them to have relevance at different scales.

In the United States, one of the most significant advances which was very much informed by the work of the United Nations came in the field of environmental justice. The sociologist Robert Bullard, often considered the father of environmental justice, published the book *Dumping in Dixie: Race Class and Environmental Quality* in 1990 (Bullard 1990). In that book, he looked at the distribution of hazardous waste dumps in the American South. He noted that the civil rights and environmental movements were converging in the late 1980s and early 1990s to confront environmental racism and promote environmental justice. Since then, many have conducted research to show that people of color and low-income communities disproportionately feel the impacts of environmental problems in the United States. This work led to the development of an environmental justice office in the Environmental Protection Agency that helps communities organize on environmental issues and that provides technical assistance.

THE DOMINANCE OF THE CITY IN THE MODERN SUSTAINABILITY MOVEMENT

As noted in the opening paragraph of this chapter, much of the work conducted on sustainability takes place in cities. To date, the city has been the place where most sustainability work occurs. This makes logical sense. Cities are highly organized entities with top down leadership. An elected official, typically a mayor, or appointed leader, such as a city manager, can set

Goal 11: Make Cities Inclusive, Safe, Resilient, and Sustainable

Target 11.1 By 2030, ensure access for all to adequate, safe and affordable housing and basic services and upgrade slums

Target 11.2 By 2030, provide access to safe, affordable, accessible and sustainable transport systems for all, improving road safety, notably by expanding public transport, with special attention to the needs of those in vulnerable situations, women, children, persons with disabilities and older persons

Target 11.3 By 2030, enhance inclusive and sustainable urbanization and capacity for participatory, integrated and sustainable human settlement planning and management in all countries

Target 11.4 Strengthen efforts to protect and safeguard the world's cultural and natural heritage

Target 11.5 By 2030, significantly reduce the number of deaths and the number of people affected and substantially decrease the direct economic losses relative to global gross domestic product caused by disasters, including water-related disasters, with a focus on protecting the poor and people in vulnerable situations

Target 11.6 By 2030, reduce the adverse per capita environmental impact of cities, including by paying special attention to air quality and municipal and other waste management

Target 11.7 By 2030, provide universal access to safe, inclusive and accessible, green and public spaces, in particular for women and children, older persons and persons with disabilities.

Target 11.A Support positive economic, social and environmental links between urban, peri-urban and rural areas by strengthening national and regional development planning

Target 11.B By 2020, substantially increase the number of cities and human settlements adopting and implementing integrated policies and plans towards inclusion, resource efficiency, mitigation and adaptation to climate change, resilience to disasters, and develop and implement, in line with the Sendai Framework for Disaster Risk Reduction 2015-2030, holistic disaster risk management at all levels

Target 11.C Support least developed countries, including through financial and technical assistance, in building sustainable and resilient buildings utilizing local materials

Source: United Nations, https://www.un.org/sustainabledevelopment/cities/

Figure 1.1. Seventeen Sustainable Development Goals (SDGs) and the goals and targets for SDG 11: Sustainable cities and communities.

an agenda for all departments that answer to him or her. If s/he decides to provide a sustainability agenda for the community s/he can easily direct agency leaders to come up with goals, targets, and agendas. There are also many professionals within a large city government to work on the goals. Thus, if a city seeks to improve access to green space or reduce greenhouse gas emissions, it is relatively easy to direct department heads to work with their staff to achieve particular outcomes. The organizational structure and professional staff present in large urban governments makes doing sustainability possible. The complex nature of cities allows them to address complex sustainability issues that cross expertise areas. Sustainability problems cross technical, social, and economic boundaries. Cities, particularly large ones, have experts that can work in teams to solve problems like greenhouse gas pollution or hurricane vulnerability with existing staff.

In addition, professionals working in city governments are well connected. There are a number of organizations, such as the United States Council of Mayors and the International City/County Management Association where experts provide expertise and networking opportunities to share best practices and emerging trends in technology. Thus, if a city manager seeks to find examples of how other cities have addressed environmental justice issues, s/he can easily contact a national network of experts to find information that will be useful. Over the years, some city governments have specialized in particular areas and have served as case studies as to how to manage difficult sustainability issues. Los Angeles, for example, has long studied how to manage vexing air pollution issues and often serves as a model for other communities.

Plus, cities have a strong tax base with funding to address complex issues. As residents and businesses have moved back into urban centers, real estate has boomed. With this growth comes an increased tax base that is helping cities ramp up sustainability initiatives. Cities can afford to invest in green energy, improved wastewater treatment plants, enhanced recycling, environmental justice, and a myriad of other projects. New York, for example, initiated a new food recycling program over the last several years (Rueb 2017) and Tampa has focused on brownfield redevelopment and environmental justice (City of Tampa 2018).

Cities also have their champions. Perhaps the biggest champion of urban sustainability initiatives is former New York mayor Michael Bloomberg (Bagley and Gallucci 2013). He pushed New York City to become more sustainable during his tenure as mayor and built national and international

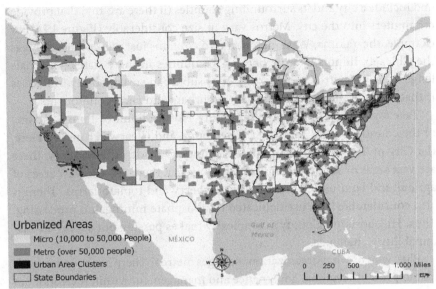

Figure 1.2. Urban/suburban areas across the United States. Urban areas/clusters (*in black*) represent urban cores, metropolitan (metro) regions include the urban core and surrounding urban areas (suburbs) with relatively higher populations than emerging suburbs (now defined as micropolitan regions). (Source: U.S. Census Bureau 2010 TIGER files for urban area clusters [UAC] and combined statistical areas [CBSA].)

connections to create networks of cities and governments. He has donated millions of dollars to sustainability initiatives and is currently leading the C40 Cities Climate Leadership Group that connects 90 global cities to try to address the causes and impacts of global climate change (Lee 2018). Of course, Bloomberg is not the only voice promoting urban sustainability. Many urbanists promote the notion of urban sustainability.

There is no doubt that there is wonderful work taking place in cities. However, as the previous paragraphs note, cities are organized to take on sustainability issues, they have the funds to address problems, and they have organizations and champions who are leading efforts to tackle tough problems.

Yet where are the suburbs in all of this? Some have tried to broaden the notion of what a city is by using the metro as a broader urban region. The metro idea, in part championed by the work of the Brookings Institute, highlights that cities are not just their jurisdictional boundaries, but also include the surrounding suburbs. Metros are defined by county boundaries and include a city and its surrounding counties (if there are any) that provide commuters into the city. Metros vary in size considerably (Figure 1.2). For example, the Yakima, Washington, metro contains just one county whereas the Chicago, Illinois, metro contains 16 counties in three states. The challenge in using metros as an organizer of suburban sustainability initiatives is that counties often contain urban, suburban, and rural components.

Take for example, Hillsborough County, Florida, home of Tampa and one of the counties in the Tampa, St. Petersburg, and Clearwater metro. Here, the City of Tampa is the core of the county. Outside of the county, there are vast areas of suburban landscapes. However, there are also vast areas of agricultural land uses, particularly strawberry and tomato farms. There is also considerable land use dedicated to phosphate mining and processing. Thus, Hillsborough County's complexity makes parsing out suburban sustainability a challenge.

Plus, who or what organization manages metros? There has been some effort to promote regional governance and management within metropolitan regions to tackle regional problems. Metropolitan planning organizations seek to find ways to coordinate planning, including sustainability planning, across jurisdictional borders. Regional economic development organizations, which are more frequently including sustainability within their missions, try to advance a regional economic agenda. Certainly the suburbs are included within these initiatives, but often these organizations do not have the teeth or political clout as their urban counterparts. Promoting sustainability within suburban landscapes is difficult within the context of the metro. While it can be done effectively, it is also a challenge and it certainly is not the norm.

Time to Focus on the Suburbs

Because suburbs have not been intentionally included in most sustainability initiatives in the United States and because they have such a strong impact on the sustainability of our nation, it is time to refocus attention on the suburbs. Spatially, the land covered by suburbs is far greater than that covered by cities. Returning to the Hillsborough County, Florida, example, one can see in Figure 1.3 that the City of Tampa is quite small compared to the entirety of the county. That is the case in most places. The spatial impacts are tremendous. The per capita density is also significantly lower in the suburbs than in the cities. The population density of Hillsborough County is about 1,400 people per square mile compared to the population density of Tampa

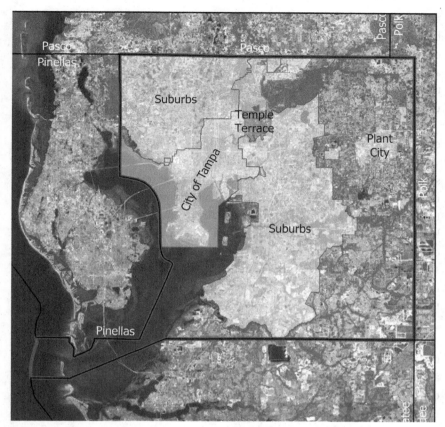

Figure 1.3. Proximity of suburban land use near the City of Tampa, Florida.

which is about 3,300 people per square mile. The infrastructure needed to support the lower density suburbs is considerable.

The suburbs are particularly challenging because they do not have strong governmental structures to support sustainability initiatives. Sometimes, the local government representing suburban interests is the county. Sometimes it is a small village government. County governments do not typically focus on small local issues of suburbs because they are broad governments seeking to address regional issues. Local villages do not have the expertise or staffing to take on complex sustainability issues. Due to these challenges, governments representing suburbs are not typically in a position to make advances. Certainly there are some wonderful exceptions, but overall local governments in suburbs can find it challenging to take on sustainability projects. Plus, many suburbs have local homeowner regulations that can make impactful changes difficult. For example, the ubiquitous suburban lawn is not an especially sustainable ground cover but is beloved by homeowners' associations. It would be difficult in many communities to advocate a more sustainable ground cover due to the power of these organizations in the management of many suburban landscapes. Some communities value a uniform look for roofs and ban rooftop solar.

Politically, leadership in suburban areas is often issue oriented. Leaders regularly emerge out of a local advocacy work. Perhaps a parent who was a good fundraiser for new marching band uniforms gets elected to a school board. Or someone who worked with local residents on a petition to improve a stoplight gets elected to a village board. While there are exceptions, political leaders in suburbs typically do not have the full breadth of knowledge to take on complex issues such as managing a polluted and declining aquifer or integrating local policies into regional, national, or global sustainability initiatives. Indeed, political leaders in the suburbs tend to be issue oriented within the local context and relatively distrustful of regional integration due to potential implications to their tax burden.

Also, suburbs often do not have the funding to invest in sustainability initiatives. People typically choose to live in the suburbs because they are cheaper than living in cities. The low tax rates found in many suburban areas attract people who, for one reason or another, do not want to pay or cannot afford to pay the type of property taxes that would support amenities or unique initiatives that would support sustainability. The services in many suburbs are lean operations that provide rudimentary services. Some suburban areas do not have sewage or drinking water systems and rely on well water and individual septic systems. Roads can be congested. Roadways have limited sidewalks and bike lanes. Mass transit is difficult due to the low density of development. The suburbs also have limited social infrastructure. Access to social services, healthcare, veterans' services, homeless shelters, drug abuse counseling, immigration services, and a myriad of other types of services and safety nets are limited. Some religious and social organizations do try to help out in some communities, but overall, the suburbs do not have the types of service institutions present in their urban counterparts.

This problem gets back to the issue raised at the outset of the book: the suburbs are changing. They are more frequently locations where the types of services found in cities are needed. For example, who would have guessed that suburban and exurban Long Island would emerge as a national focus for the problem with the Central American gang MS-13. The gang is responsible for a number of murders on the island and terrorizes some suburban neighborhoods. In 1957, the Tony Award–winning musical *West Side Story* highlighted the problems between gangs in the West Side of Manhattan. Today, that community is gentrified and the heart of a thriving LGBTQ neighborhood. The gang problem is now a suburban issue. Sam Quinones captured the changing suburbs in parts of the Midwest and Appalachia in his groundbreaking book *Dreamland* which detailed the opioid epidemic occurring in the suburbs (Quinones 2015). The problems demonstrated by the advent of MS-13 and the suburban opioid and designer drug problem show that the suburbs are not equipped to address the modern problems that are emerging in society. Plus, with the existential challenges faced in some suburbs due to these types of problems, it makes it difficult to get sustainability on a local agenda in a community with limited resources.

Plus, in many areas of the country, the housing stock of the suburbs is getting older. The first mass-produced houses were built in the 1940s and 1950s. Since then, the suburbs have expanded considerably. The oldest communities are over 70 years old. Many of them are showing their age. Some have gone through continuous redevelopment and continue to thrive. Levittown, New York, for example, remains a desirable neighborhood. However, others have seen considerable decline and reduction of property values. Of course, the decline leads to a loss of tax revenue and even less funding for infrastructure, schools, and services.

Even with all of these issues, the suburbs remain a draw for many. The low cost of living is certainly a strong pull. However, the low-density housing, the newer housing stock, and the perceived ease of lifestyle continue to draw new residents. Throughout the United States, one can find thriving suburban communities of all ages. New suburban communities are being built to try to alleviate some of the infrastructure problems of older suburbs by enhancing density, addressing mass transit, enhancing drinking water and sewage systems, and adding sustainability measures. Some older suburbs are undergoing a redevelopment phase by infilling for density, enhancing infrastructure, and confronting sustainability issues. The suburbs are getting a second look as developers try to make them more attractive to a new generation that is more interested in urban living. Plus, despite some of the problems that were identified in the previous paragraphs, there are leaders emerging who are interested in addressing the myriad of challenges present in today's modern suburbs.

Christopher Sellers points out in his groundbreaking book, *Crabgrass Crucible: Suburban Nature and the Rise of Environmentalism in Twentieth-Century America*, the suburbs have long been the crucible for many environmental initiatives that emerged in the United States (Sellers 2012). While they have taken a bit of a back seat in recent decades as the cities led the charge on global sustainability initiatives, this book demonstrates that the suburbs are taking a more important position as the country grapples with many of the sustainability challenges it is facing.

Book Organization

Much of the discourse surrounding sustainability is compartmentalized into thematic topics like the three *e*'s of sustainability (environment, economy, and society), and books on sustainability are often organized by these general pillars (e.g., we did this in a different book that examines case studies from an international view of sustainability [Brinkmann and Garren 2018]). Many scholars have expanded the three-pillar model to include additional pillars. For example, Sachs (2015) added "good governance" as a fourth pillar which includes leadership, policy development, and coordinated strategies between multiple levels of government. The book generally follows this convention; however, has been adapted to incorporate a few key things that have been emerging in the field.

In this book, we pull together 15 cases from across the United States (Figure 1.4) to provide a robust and comprehensive collection of suburban sustainability stories. Given the continued theme of strong leadership as demonstrated through sustainability planning and policy development, we begin our consideration of cases through the lens of planning in part I. Part II focuses on two of the pillars of sustainability, social and economic issues. Part III contains a focused look at two of the key concerns many suburban communities are facing, air pollution and climate change. Finally, part IV contains several case studies in the area of sustainable land and water management. Given the interrelated quality of sustainability, Table 1.1 provides highlights from each chapter.

Taken together, this is the largest collection of case studies focused on suburban sustainability issues. While this book is not meant to be comprehensive spatially or thematically, it does highlight many of the themes associated with sustainability in suburban America. We hope it will be useful for providing a context for greater sustainability research in the coming years in suburban settings. As this book demonstrates, there is a strong need for making our suburbs more sustainable—and it is possible.

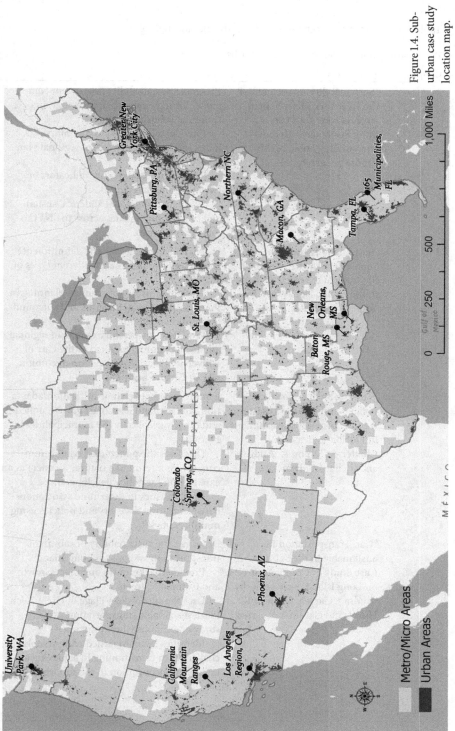

Figure 1.4. Sub-
urban case study
location map.

Table 1.1. Suburban sustainability case study chapter highlights

Part I. Sustainability Planning in the Suburbs

Chapter	Title and Author	Highlights
2	"Sustainability Assessment of the New York Metropolitan Area: Is New York City More Sustainable Than Its Surrounding Suburbs?" by Sandra J. Garren	Presents a sustainability assessment completed for regional, urban core, and surrounding suburban counties. Provides a balanced sustainability index methodology based on regional plan objectives. Compares sustainability indicators between regions and counties. Summarizes regional and local sustainability planning policies across the NYC metropolitan area.
3	"Sustainability through Suburban Growth? Regional Planning and Targeted Change in the Pacific Northwest" by Yonn Dierwechter	Discusses the nuances of the definition of suburbs and suburbia and different types of suburbs. Presents a history of regional planning in Puget Sound, Greater Seattle, and surrounding suburban regions. Provides a successful example of regional and local planning targeted in the City of University Place, WA (a suburban community in Greater Seattle).
4	"Linking Smart Growth Policies and Natural Disasters: A Study of Local Governments in Florida" by Vaswati Chatterjee, Simon A. Andrew, and Richard C. Feiock	Examines climate change mitigation/adaptation policies, land use/smart growth policies, and other variables in 165 municipalities in Florida. Discusses the punctuated equilibrium concept (as a result of a natural disaster) as an explanation for policy development. Uses statistics to determine associations between natural disasters and policies using multiple data sets.
5	"Can Transportation Be Sustainable in the Suburbs? A Case Study of Complete Streets on Long Island, New York" by Sandra J. Garren, Mathew K. Huxel, and Carolina A. Urrea	Examines the sustainability of suburban transportation systems on Long Island. Provides background on Smart Growth and Complete Streets. Uses the success of Village of Great Neck Plaza sustainable transportation as a case study of exemplary best practices. Identifies lessons learned and presents challenges to transitioning to achieving sustainable transportation in the suburbs.

Part II. Socioeconomic Sustainable Development in the Suburbs

Chapter	Title and Author	Highlights
6	"Who Defines Sustainability? Urban Infill in Colorado Springs, Colorado" by John Harner	Tells the story of a commercial development next to a university that was supposed to be a sustainable development project but ended up being a sterile place.
		Discusses the challenges between private land use development concerns with sustainable development practices.
		Provides an example of partnerships between universities, land developers, and local government.
7	"Diversity Improves Design: Sustainable Place-Making in a Suburban Tampa Bay Brownfield Neighborhood" by E. Christian Wells, Gabrielle R. Lehigh, Sarah Combs, and Miles Ballogg	Presents a success story of redevelopment in a marginalized community in Tampa, Florida.
		Discusses the difference between "Space versus Place," "embedded versus disembedded approaches," and "Brownfields versus "Healthfields" in community development.
		Provides a history of Brownfield and redevelopment into a "Healthfield."
		Provides results from long history of engagement (through survey and interviews from authors) from local university.
8	"Sustainable Economic Development: The Maker Movement in Macon, Georgia" by Susan M. Opp	Tells the story of how an entrepreneur program (i.e., SparkMacon) has advanced sustainability principles.
		Provides a detailed background on the "Maker Movement."
		Highlights the benefits to sustainability in local areas that avoids negative externalities and increases human development opportunities.

Part III. Focus on Atmospheric Sustainability in the Suburbs

Chapter	Title and Author	Highlights
9	"Ozone in Urban North Carolina: A Sustainability Case Study" by William H. Battye, Casey D. Bray, Pornpan Uttamang, and Viney P. Aneja	Provides a success story on how this suburban region attained ozone compliance.
		Details the history of ground-level ozone and regulations to address it at the federal (i.e., Clean Air Act) and state, and local level in North Carolina.
		Documents the history of Ozone Concentrations in North Carolina and within the region.

(continued)

Table 1.1—*Continued*

Chapter	Title and Author	Highlights
10	"Skewed Sustainability and Environmental Injustice across Metropolitan St. Louis, Missouri" by Troy D. Abel, Stacy Clauson, and Debra Salazar	Tells the story of how St. Louis is addressing sustainability to include consideration of environmental justice issues with some success, but challenges remain (e.g., continued exposure to air pollution). Documents the planning process in the city (i.e., OneSTL). Discusses how regional planning with federal funding helped develop OneSTL.
11	"Suburban Sustainability Governance in the Los Angeles Region, California" by Elizabeth Mattiuzzi	Presents a successful policy regime at the subregional scale in the Los Angeles region (Los Angeles, San Bernardino, and Riverside Counties). Examines the variation in local governments within the same regional and state policy environment. Provides a summary of state climate policy and land use policy across the region.
12	"The Application of Land Use Regression and the National Land Cover Dataset in Modeling of Ozone Mixing Ratios in Baton Rouge, Louisiana" by Mallory Thomas	Provides background on ozone and effects and movement patters of ozone. Presents a method (i.e., regression analysis) to analyze ozone mixing patters across different land cover to see how mixing occurs in the Baton Rouge areas.

Part IV. Sustainable Land and Water Management in the Suburbs

Chapter	Title and Author	Highlights
13	"Leading through Water: Defining Sustainability through Leadership, Experience, and Engagement in Pittsburgh, Metropolitan Region, Pennsylvania" by Michael H. Finewood and Sean McGreevey	Provides a discussion on the nuances with the broad definition and tools used in the field of sustainability education. Summarizes a case study on a class that did experiential practice in Pittsburgh that involved conceptualizing sustainability. Demonstrated that students had positive experiences with engaging with nature and real-world sustainability challenges.
14	"Managing Wildlife amid Development: A Case Study of Sustaining Mountain Lion Populations in California" By Melissa M. Grigione, Michaela C. Peterson, Ronald Sarno, and Mike Johnson	Provides a case of how wildlife (mountain lions in this case) can successfully share land with people. Studied the range of mountain lions in suburban California. Mountain lions in suburban areas may be subject to inbreeding due to the limited range; thus corridors are key to long-term population viability.

Chapter	Title and Author	Highlights
15	"Suburban Unsustainability or the Burden of Fixed Infrastructure in New Orleans, Louisiana" by Craig E. Colten	Tells the story of floodwater infrastructure (i.e., levees and pumps) that failed in the New Orleans metro region. Reviews challenges with suburban sustainability in the context of a hazardous landscape. Suggests that suburbs can be more vulnerable than cities unless they are developed with hazards in mind.
16	"Suburban to Urban Hydro-Economic Connectivity: Virtual Water Flow within the Phoenix Metropolitan Area, Arizona" by Richard R. Rushforth and Benjamin L. Ruddell	Provides a method to account for water flow within the Phoenix metro region. Summarizes water supply chain issues within an urban/suburban regional context. Demonstrates a model by which water planers can contextualize water uses of core cities and peripheral areas.

References Cited

Anacker, Katrin, Christopher Niedt, and Chang Kwon. 2017. "Analyzing Segregation in Mature and Developing Suburbs in the United States." *Journal of Urban Affairs* 36(6): 819–832.

Bagley, Katherine, and Maria Gallucci. 2013. "New York's New Mayor Must Build on Michael Bloomberg's Green Legacy." *The Guardian New York Environment Blog* December 10.

Boustan, Leah Platt. 2010. "Was Postwar Suburbanization 'White Flight'? Evidence from the Black Migration." *The Quarterly Journal of Economics* 125(1): 417–443.

Brinkmann, Robert, and Sandra Garren. 2018. *The Palgrave Handbook of Sustainability: Case Studies and Practical Solutions*. Basingstoke, United Kingdom: Palgrave Macmillan.

Brueckner, Jan K. 2000. "Urban Sprawl: Diagnosis and Remedies." *International Regional Science Review* 23(2): 160–171.

Brundtland Commission. 1987. *Our Common Future*. Oxford: Oxford University Press.

Bullard, Robert D. 1990. *Dumping in Dixie: Race, Class, and Environmental Quality*. Boulder, Colorado: Westview Press.

Carson, Rachel. 1962. *Silent Spring*. Boston: Houghton Mifflin.

City of Tampa. 2018. "Brownfields Program." Last accessed June 2, 2018. https://www.tampagov.net/economic-and-urban-development/programs/brownfields-assessment-grant-program.

Duany, Jorge. 2011. *Blurred Borders: Transnational Migration Between the Hispanic Caribbean and the United States*. Chapel Hill: University of North Carolina Press.

Foner, Nancy. 2000. *From Ellis Island to JFK: New York's Two Great Waves of Immigration*. New Haven: Yale University Press.

Gibbs, Lois Marie. 2011. *Love Canal and the Birth of the Environmental Health Movement.* Washington, D.C.: Island Press.

Hardin, Garrett. 1968. "The Tragedy of the Commons." *Science* 162(3859): 1243–1248.

Higgins, Richard. 2017. *Thoreau and the Language of Trees.* Berkeley: University of California Press.

Hyra, Derek. 2015. "The Back-to-the-City Movement: Neighbourhood Redevelopment and Processes of Political and Cultural Displacement." *Urban Studies* 52(10): 1753–1773.

Kahn, Matthew E. 2000. "The Environmental Impact of Suburbanization." *Policy Analysis Management* 19(4): 569–586.

Lee, Taedong. 2018. "Network Comparison of Socialization, Learning, and Collaboration in the C40 Cities Climate Group." *Journal of Environmental Policy and Planning.* DOI: 10.1080/1523908X.2018.1433998.

Leopold, Aldo. 1949. *A Sand County Almanac.* Oxford: Oxford University Press.

Murdoch, James C., and Todd Sandler. 1997. "The Voluntary Provision of a Pure Public Good: The Case of Reduced CFC Emissions and the Montreal Protocol." *Journal of Public Economics* 63(3): 331–349.

Myers, Dowell. 2016. "Peak Millennials: Three Reinforcing Cycles That Amplify the Rise and Fall of Urban Concentration by Millennials." *Housing Policy Debate* 26(6): 928–947.

Niedt, Christopher. 2013. *Social Justice in Diverse Suburbs: History, Politics, and Prospects.* Philadelphia: Temple University Press.

Novak, Barbara. 2007. *American Painting of the Nineteenth Century: Realism, Idealism, and the American Experience.* 3rd ed. Oxford: Oxford University Press.

Quinones, Sam. 2015. *Dreamland: The True Tale of America's Opiate Epidemic.* London: Bloomsbury.

Righter, Robert W. 2005. *The Battle over Hetch Hetchy: America's Most Controversial Dam and the Birth of Modern Environmentalism.* Oxford: Oxford University Press.

Roseland, Mark. 2012. *Toward Sustainable Communities: Solutions for Citizens and Their Governments.* 4th ed. British Columbia: New Society Publishers.

Rueb, Emily S. 2017. "How New York Is Turning Food Waste into Compost and Gas." *New York Times* June 2.

Sachs, Jeffrey D. 2015. *The Age of Sustainable Development.* New York: Columbia University Press.

Schneider, Jack. 2008. "Escape from Los Angeles: White Flight from Los Angeles and Its Schools, 1960–1980." *Journal of Urban History* 34(6): 995–1012.

Sellers, Christopher C. 2012. *Crabgrass Crucible and the Rise of Environmentalism in Twentieth-Century America.* Chapel Hill: University of North Carolina Press.

Short, John Rennie, Bernadette Hanlon, and Thomas J. Vicino. 2007. "The Decline of Inner Suburbs: The New Suburban Gothic in the United States." *Geography Compass* 1(3): 641–656.

Soja, Edward W. 2014. *My Los Angeles: From Urban Restructuring to Regional Urbanization.* Berkeley: University of California Press.

Strom, Elizabeth. 2017. "How Place Matters: A View from the Sunbelt." *Urban Affairs Review* 53(1): 197–209.

United Nations Environment Programme. 2018. "The Montreal Protocol on Substances that Deplete the Ozone Layer." Last accessed March 5, 2020. https://web.archive.org/web/20130602153542/http://ozone.unep.org/new_site/en/montreal_protocol.php.

Walker, K. E. 2014. "Immigration, Local Policy, and National Identity in the Suburban United States." *Urban Geography* 35(4): 508–529.

I

Sustainability Planning
in the Suburbs

2

Sustainability Assessment of the New York Metropolitan Area

Is New York City More Sustainable Than Its Surrounding Suburbs?

SANDRA J. GARREN

There has been a long-standing debate as to whether urban areas are more sustainable than the suburbs (Alexe 2016; Green 2018). Suburbs have generally been deemed less sustainable due to the higher energy consumption caused from sprawling homes and a car-dependent culture (Glaeser 2009). However, quality of life and economic factors are often not included in this discussion, which, along with the environment, are basic tenets of sustainability. This chapter presents a sustainability index that balances the values (i.e., equity, environment, health, and prosperity) of the most recent regional plan and quantitative results to assess the sustainability of the urban core and the surrounding suburbs. A review of sustainability efforts made in each sub-region is also presented.

Sustainable development was popularized through the Brundtland Commission in the late 1980s and was officially defined as "development that meets the needs of the present without compromising the ability of future generations to meet their own needs" (World Commission on Environment and Development 1987). Intrinsic to the definition is the idea of preserving the environment. In addition to clean air, water, and energy, it is paramount to also meet the needs of people which means that we must ensure that everyone has equal opportunity to medical care, healthy food, decent jobs, and affordable housing.

Measuring progress toward sustainable development (or sustainability as will be used interchangeably in this chapter) from the vague Brundtland definition has proven to be problematic since there is no agreed upon formula to use to assess progress. In attempts to measure progress, governments, businesses, and academic researchers have crafted sustainability assessment methodologies, and there is no shortage of sustainability assessment tools. For example, Singh et al. (2009) identified 72 unique sustainability assessment indices that range from application in governments to industries (e.g., Human Development Index, City Development Index, Composite Sustainable Development Index for Industries, and Environmental Quality Index). The general sustainability assessment process entails creating a framework, identifying indicators, and developing indices for benchmarking purposes. These indices have become useful tools in informing policy and for communicating sustainability to the public (Singh et al. 2009).

Sustainability assessments can be conducted at any scale (e.g., nations, states, cities, and businesses) and can be conducted using top-down or bottom-up approaches. A recent trend is to incorporate the newly developed United Nation's global sustainable development goals (SDGs) into the assessment framework. For example, New York City has begun to incorporate global goals in its sustainability plans by matching up their initiatives with the SDGs (New York City 2015a). In 2016, 17 SDGs were agreed upon and came into effect that include social parameters such as gender equity, health, and education; environmental issues such as climate change, energy, and water supply); and, economic issues such as poverty and access to jobs (United Nations Development Program undated). The SDGs have a combined total of 169 targets and 232 unique indicators to assess national progress toward sustainable development (United Nations 2018). The United Nations has created a global database to track these indicators (United Nations Statistics undated) and in 2017, a global SDG Index was developed to benchmark and compare world nations and annual assessments have continued through 2019 (Bertelsmann Stiftung and Sustainable Development Solutions Network undated).

Sustainability assessment at the sub-national level (e.g., regions, states, counties, and cities) is not as well developed and generally focuses on large cities which leaves out surrounding suburbs and exurbs. Some attempts have been made to assess sustainability in large metropolitan areas. The United States (U.S.) Cities SDG Index is a notable example of assessment at the metropolitan scale (Bertelsmann Stiftung and Sustainable Development Solutions Network undated).

This chapter attempts to fill this void by presenting a regional sustainability assessment framework and sustainability index that compares the urban core with the surrounding suburbs in the New York metropolitan area. The framework and index presented in this chapter are aggregated by the counties that represent the urban core (defined as New York City) and suburban counties (defined as the suburbs and exurbs in the surrounding counties of New York City). From the results of the analysis, it is possible to quantitatively compare the regions and counties within the metro area. The chapter also inventories regional sustainability planning efforts and identifies specific advocacy groups in the study area that could influence progress toward sustainable development.

CASE STUDY: NEW YORK METROPOLITAN AREA

The New York metropolitan area is home to nearly 23 million people located in three states, five sub-regions (New York City, Long Island, Hudson Valley, Northern New Jersey, and Southwest Connecticut), and 31 counties (Figure 2.1; Table 2.1). Within the area, there are 975 local governments that include boroughs, towns, cities, villages, and unincorporated census designated places (CDPs). Land use is diverse across the region and contains 16 different land use typologies organized by the Regional Plan Association (RPA) into five general groupings that contain the urban core, commercial-industrial, primarily residential, downtown and centers, and rural and open spaces (Regional Plan Association 2015)(Figure 2.2).

The urban core in this study is defined as New York City (NYC) which contains five counties which are also known as boroughs (i.e., New York County [Manhattan Borough], Queens County [Queens Borough], Kings County [Brooklyn Borough], Bronx County [the Bronx Borough], and Richmond County [Staten Island Borough]). The surrounding counties are primarily suburban residential, which in this study is defined as counties with population densities greater than 1,000 people per square mile. Beyond the adjacent more urbanized suburban counties are a number of exurban communities and open spaces. The Regional Plan Association's "downtown and centers" land use grouping (shown in black on Figure 2.2) depicts the many urban areas that have grown up around the urban core that have become major cities in their own right, such as Jersey City in Northern New Jersey which has an urban character and a high population density (nearly 15,000 persons per square foot). As shown on Table 2.1, the population in counties in the metro area ranges from as few as 76,330 people in Sullivan County, New York, to 2.3 million people in Queens County, New York (the most populated county in the metro area).

Table 2.1. Population and population densities in New York metropolitan area counties and sub-regions

SOUTHWEST CONNECTICUT

Fairfield 939,983 (1,116)
Litchfield 186,304 (199)
New Haven 862,224 (1,001)
Total 1,988,511

NORTHERN NEW JERSEY

Bergen 926,330 (4,028)
Essex 791,609 (6,318)
Hudson 662,619 (14,609)
Hunterdon 126,250 (293)
Mercer 370,212 (1,654)
Middlesex 830,300 (2,722)
Monmouth 629,185 (1,341)
Morris 498,192 (1,086)
Ocean 583,450 (936)
Passaic 507,574 (2,768)
Somerset 330,604 (1,106)
Sussex 145,930 (277)
Union 548,744 (5,404)
Warren 107,226 (299)
Total 7,058,225

HUDSON VALLEY, NEW YORK

Dutchess 296,928 (371)
Orange 375,384 (465)
Putnam 99,488 (430)
Rockland 320,688 (1,878)
Sullivan 76,330 (77)
Ulster 181,300 (160)
Westchester 967,315 (2,268)
Total 2,317,433

LONG ISLAND, NEW YORK

Nassau 1,354,612 (4,781)
Suffolk 1,501,372 (1,646)
Total 2,855,984

NEW YORK CITY, NEW YORK

Bronx 1,428,357 (34,571)
Kings 1,595,259 (37,232)
New York 1,629,507 (72,033)
Queens 2,301,139 (21,553)
Richmond 472,481 (8,130)
Total 7,426,743

NEW YORK METROPOLITAN AREA 2015 POPULATION WAS 22,464,897

Notes: The 2015 population data was obtained from the United States Census. Population density is reported in people per square mile.

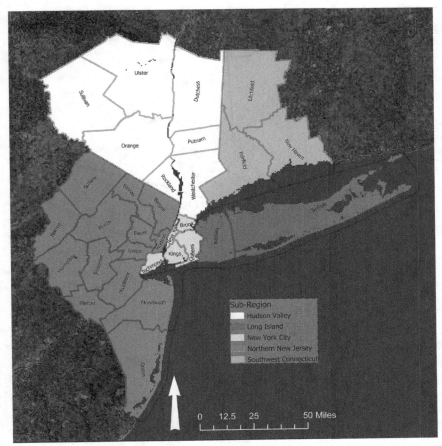

Figure 2.1. Sub-regions and counties in the Greater New York metropolitan area.

Sustainability Index Methodology

To compare urban and suburban counties in the New York metropolitan area, a sustainability index was developed within the five sub-regions and 31 counties under the purview of the Regional Plan Association (RPA). The RPA is a research and planning organization that has been guiding the planning process in the New York-New Jersey-Connecticut (NY-NJ-CT) metropolitan area for more than 90 years. The most recent regional plan, Charting a New Course: A Vision for a Successful Region, identifies current trends, lays out scenarios for future growth, and provides an aspirational vision for 2040 (Regional Plan Association 2017). The aspirational vision contains four main elements. The first is to enhance prosperity and provide opportunity for all people in the region. The vision focuses on placing jobs

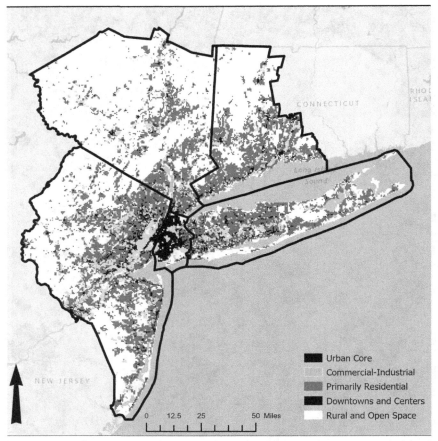

Figure 2.2. Land use groups in the New York metropolitan area. (Source: Recreated from data obtained from Regional Plan Association 2016.)

in mixed-use developments near public transportation and increasing the number of jobs for low-income workers. The second vision is to improve the affordability of housing, particularly in areas with low vacancy rates and high housing costs. It also focuses on reducing commute times. The third vision is to foster a healthier and more livable environment by increasing mixed-use development close to fresh food, education, and access to nature. It is widely known that building walkable communities improves air quality and health. And lastly, the fourth vision is to improve the resiliency of the region through higher density developments that would reduce greenhouse gas emissions. In the RPA plan, new growth would be prohibited in areas prone to flooding or in environmentally sensitive areas.

In addition to the aspiration vision, four values are identified in the RPA plan as follows:

Equity. The elimination of poverty and homelessness were identified as core values well as closing health and wealth gaps.

Health. The vision is to provide access to clean air, healthy food, and health care for all residents in the region.

Environmental Sustainability. The plan values the environment and has specified goals to reduce greenhouse gas (GHG) emissions by 80 percent by the year 2040, eliminate raw sewage discharges, and improve the resiliency to the negative effects from climate change (i.e., flooding and extreme heat).

Prosperity. The 2040 plan lays out a goal to create millions of jobs, increase wages for all, and focus on improving life for the poor.

The four values designated by the RPA (i.e., Equity, Health, Environment, and Prosperity) were selected to form the framework for the sustainability index for this analysis. The next step in the sustainability assessment process was to identify a list of potential sustainability indicators from the academic literature and from state, regional, and local government metrics. Standard indicators were evaluated from the previously mentioned U.S. Cities SDG Index which ranked 100 American metropolitan areas, including the New York metropolitan area (Sustainable Development Solutions Network 2017). The U.S. Cities SDG Index included 49 indicators that gave equal weight to indicators grouped into 16 of the 17 SDGs (Sustainable Development Solutions Network 2017). While many of these indicators are relevant with data available at the county level, some of the data were not available or applicable at the county level, and organizing the framework by SDGs would provide too many gaps to be useful. In addition, the 55 indicators used to measure progress in New York City's OneNYC plan were also evaluated for inclusion in the larger metro area (New York City 2014). These indicators are organized around their four vision areas (i.e., Our Growing Thriving City, Our Just and Equitable City, Our Sustainable City, and Our Resilient City). As might be expected, many of the indicators would not have data at the county level in areas outside New York City.

For this index, 20 indicators were selected based on relevancy and readily available data aggregated at the county level (Figure 2.3). Four to six indicators were identified for each value (i.e., equity, health, environment, and prosperity) and are presented in Tables 2.2 through 2.5, respectively, along with justification for inclusion in the framework. This grouping of

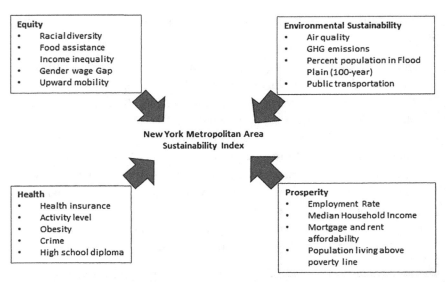

Equity
- Racial diversity
- Food assistance
- Income inequality
- Gender wage Gap
- Upward mobility

Environmental Sustainability
- Air quality
- GHG emissions
- Percent population in Flood Plain (100-year)
- Public transportation

New York Metropolitan Area Sustainability Index

Health
- Health insurance
- Activity level
- Obesity
- Crime
- High school diploma

Prosperity
- Employment Rate
- Median Household Income
- Mortgage and rent affordability
- Population living above poverty line

Figure 2.3. Sustainability index framework and indicators for the New York metropolitan area.

indicators provides for a more balanced assessment of sustainability which is suggested by Komeily and Srinivasan (2015) and Happio (2012) to be a more complete analysis of sustainability. It was noted by both studies that existing sustainability frameworks (e.g., LEED for Neighborhood Development, CASBEE for Urban Communities, and BREEAM Communities) predominantly consider environmental criteria and leave out quality of life categories.

The next step in creating an index was to normalize and weight the indicators into each of the categories. Indicators that were not already in the correct percent format ranging from 0 to 100 percent were normalized using the minimum/maximum method as described in Organization for Economic Cooperation and Development (2008). To equally weight each of the four value areas, a simple arithmetic mean was calculated within each category to arrive at an overall score per category. Finally, an arithmetic mean was used to calculate the overall sustainability score for each county. Counties could then be compared using the rankings of the resulting scores. An average of counties with each of the five sub-regions was used to compare sub-regions, and an average of five counties within New York City and all the remaining counties were averaged to compare the urban and suburban overall score.

Table 2.2. Equity indicators and summary statistics for counties in the New York metropolitan area

Indicator	Dataset Description and Processing	Summary Statistics Average (Range)
Racial diversity[1]	Calculated by taking the sum of the squares of percent Hispanic and non-Hispanic white, black, and Asian populations. Higher number is more diverse.	Raw: 51 (18 to 76) (Not Scaled)
Percent population not using SNAP benefits[1]	Percent of SNAP recipients subtracted from 100.	Raw: 89% (63 to 97%) (Not Scaled)
Gender wage gap[1]	Percent difference calculated by subtracting median income of women from men relative to men. Total subtracted by 100%. The higher the number, the more equal the income is between men and women.	Raw: 61% (0 to 100%) (Not Scaled)
Income equality[1] (from Gini coefficient of income inequality)	Indexed value reported by the census as a ratio of inequality. Raw value was subtracted from 1 and multiplied by 100. Higher values represent more equal income.	Raw: 47 (41 to 60) (Not scaled)
Upward mobility[2]	Expected rank of children whose parents are at the 25th percentile of the national income distribution based on rank-rank regression. The higher the number, the greater the chance of upward mobility.	Raw: 45 (39 to 52) (Not scaled)
	Average and range of combined equity indicators	**59 (42 to 65)**

Sources: 1. Data obtained from U.S. Census, American Community Survey (2011–2015).
2. Data obtained from the Equality of Opportunity Project (Chetty et al. 2014).

Table 2.3. Health indicators and summary statistics for counties in the New York metropolitan area

Indicator	Dataset Description and Processing	Summary Statistics Average (Range)
Health insurance[1]	Civilian noninstitutionalized population—with health insurance coverage	Raw: 90% (81 to 96%) (Not scaled)
Activity level[2]	Percent of population who is not active during leisure time, subtracted by 100 to arrive at active people.	Raw: 78% (69 to 83%) (Not scaled)

(continued)

Table 2.3—*Continued*

Indicator	Dataset Description and Processing	Summary Statistics Average (Range)
Not obese[2]	Percent population who are obese. Subtracted by 100 to arrive at not-obese percentage.	Raw: 75% (70 to 85%) (Not Scaled)
Crime free[3]	Total number of violent crimes per 100,000. Scaled using min-max method for descending order. A higher number represents less crime.	Raw: 276 (40 to 681) Scaled: 63 (0 to 100)
High school diploma[1]	Percent population with high school diploma	Raw: 88% (71 to 94%) (Not Scaled)
	Average and range of combined health and well-being indicators	**79 (61 to 90)**

Sources: 1. Data obtained from U.S. Census, American Community Survey (2011–2015).
2. Data obtained from the CDC.
3. Data obtained from the FBI Uniform Crime Statistics.

Table 2.4. Environmental sustainability indicators and summary statistics for counties in the New York metropolitan area

Indicator	Dataset Description and Processing	Summary Statistics Average and Range
Good air quality[1]	EPA's air quality index (AQI) annualizes air quality data by county. The percent number of days of good air quality was calculated using the number of AQI reporting days and the number of good air quality days. Four counties did not have data, so an average of surrounding counties was used as a proxy.	75% (56 to 94%) (Not Scaled)
Carbon footprint[2]	Total carbon footprint per household per year in tons of carbon dioxide equivalents (tCO_2e). Raw data converted to positive scale. Higher number represents lower carbon footprint.	Raw: 50 (33 to 64) Scaled: 44 (0 to 100)
Population out of floodplain[3]	The dataset contains the population and household units that are in a 100-year floodplain. The percent population in the 100-year floodplain was subtracted by 100%.	95% (85 to 100%) (Not Scaled)
Public transit[4]	Percent of workers that take public transit to work (all forms of public transit). The higher number indicates more people taking public transit.	Raw: 16% (1 to 62%) (Not scaled)

(continued)

Indicator	Dataset Description and Processing	Summary Statistics Average and Range
	Average and range of combined environmental indicators	**(45 to 83)**

Sources: 1. Data obtained from Environmental Protection Agency.
2. Data obtained from Cool Air Cool Planet.
3. Data obtained from NYC Furman Center, U.S. FloodZone Map.
4. Data obtained from U.S. Census, American Community Survey (2011–2015).

Table 2.5. Prosperity indicators and summary statistics for counties in the New York metropolitan area

Indicator	Dataset Description and Processing	Summary Statistics Average (Range)
Employment rate (2015)[1]	Original data is annualized by Bureau of Labor Statistics. Employment rate calculated by subtracting from 100%.	Raw: 95% (92 to 96%) (Not scaled)
Median household income (2015)[2]	Median household income	Raw: $74,243 ($34,299 to $105,444) Scaled: 56 (0 to 100)
Median home value (2015)[2]	Median home value of owner-occupied homes	$368,339 ($165,900 to $848,700) 30 (0 to 100)
Mortgage affordability (2015)[2]	Selected monthly owner costs as a percentage of monthly owner costs as a percentage of household income (SMOCAPI). The value represents the percent population that spends less than 35.0% on housing.	66% (54 to 78%) (Not scaled)
Rent affordability (2015)[2]	Gross rent as a percentage of household income (GRAPI). The value represents the percent population that spends less than 35.0% on housing.	54% (46 to 64%) (Not scaled)
Population living above poverty level	The percent population living below poverty was subtracted from 100 to convert to a positive scale	88% (69 to 96%) (Not scaled)
	Average and range of combined prosperity indicators	**65 (49 to 79)**

Source: 1. Data obtained from the Bureau of Labor Statistics.
2. Data obtained from the U.S. Census, American Community Survey (2011–2015).

SUSTAINABILITY INDEX RESULTS

Table 2.6 provides a summary of sustainability scores by county and the scores are depicted on Figure 2.4. Averages and ranges of individual indicators were also calculated and included in the indicator lists presented in Tables 2.2 through Table 2.5 in the previous section. In all tabulations, the higher the calculated score, the more sustainable the region is as compared to lower scores in the region.

From the results, the urban core of New York City was found to be the most sustainable on all fronts. When examining the sub-regions, the average sustainability score of the five NYC counties (or Boroughs) was 67.5, followed by Long Island (66.9), Hudson Valley (66.2), and Northern New Jersey (65.7). Southwest Connecticut was lower than NYC with a score of

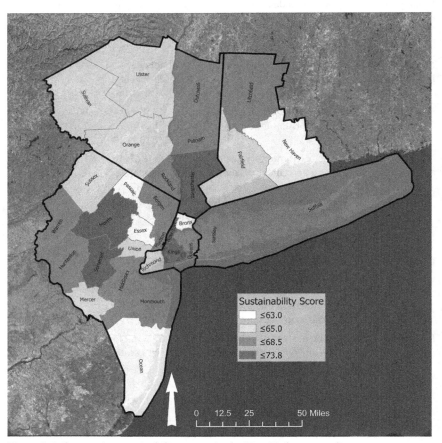

Figure 2.4. Composite sustainability scores in the New York metropolitan area.

Table 2.6. New York metropolitan area composite sustainability index summary

Sub-Region	County	Equity	Health	Environment	Prosperity	Composite Sustainability Index
Southwest Connecticut	Fairfield	63.4	80.8	46.2	69.3	64.9
	Litchfield	56.3	86.1	54.1	64.4	65.2
	New Haven	61.8	76.1	51.8	59.0	62.2
Northern New Jersey	Bergen	66.6	86.0	49.1	70.5	68.1
	Essex	66.4	62.7	64.2	57.7	62.7
	Hudson	68.0	71.5	68.8	59.8	67.0
	Hunterdon	57.8	90.2	45.4	75.9	67.3
	Mercer	65.3	74.9	53.5	63.7	64.3
	Middlesex	66.2	82.0	56.4	67.1	67.9
	Monmouth	59.0	84.6	51.5	68.3	65.9
	Morris	61.5	88.6	50.3	76.1	69.1
	Ocean	56.3	83.8	53.7	57.8	62.9
	Passaic	63.3	70.6	57.9	57.3	62.3
	Somerset	64.9	88.4	46.4	75.3	68.8
	Sussex	55.8	87.2	49.5	66.3	64.7
	Union	67.5	74.5	52.1	62.6	64.2
	Warren	57.7	85.0	53.9	63.8	65.1
Hudson Valley	Dutchess	60.1	81.7	59.5	62.6	65.9
	Orange	60.7	78.8	58.1	60.9	64.6
	Putnam	59.1	87.8	54.4	71.0	68.1
	Rockland	64.0	83.2	55.5	66.5	67.3
	Sullivan	59.2	78.6	61.9	53.7	63.3
	Ulster	58.0	82.6	60.0	57.3	64.5
	Westchester	66.1	81.5	60.1	71.0	69.7
Long Island	Nassau	64.5	83.3	52.1	73.0	68.2
	Suffolk	61.7	84.4	48.7	67.9	65.7
New York City	Bronx	62.2	61.0	78.8	49.3	62.8
	Kings	67.2	65.6	83.0	59.6	68.9
	New York	66.9	71.0	78.4	78.8	73.8
	Queens	70.0	64.5	74.5	60.5	67.4
	Richmond	62.7	66.7	63.2	65.6	64.6

64.1. In addition, when compared the urban core of the city to all other suburban counties, New York City was also found to be significantly more sustainable (67.5 in the five counties that comprise New York City compared to 65.8 in the suburbs). Lastly, when comparing counties, New York County (Manhattan Borough) was significantly more sustainable (with an overall sustainability score of 73.8), followed by Westchester, Morris, Kings, and Somerset Counties.

While overall, New York City and Manhattan had the highest sustainability scores, some individual counties and sub-regions performed close to or better than the urban area in each of the four values (Figure 2.5). For example, Hunterdon County in New Jersey took the top spot in the overall health category. Long Island was second when considering sub-regions. Westchester County in the Hudson Valley and Morris County in New Jersey took second and third spot overall and were both ahead of four of the five New York City counties in some individual categories. Southwest Connecticut scored in the middle of the pack on all fronts. Showing the geospatial distribution of scores for each category clearly shows that there are differences. For example, the urban core shows stronger sustainability in equity and the environment, while the suburbs are stronger in health and prosperity.

POLICIES AND PROGRAMS IN THE NEW YORK METROPOLITAN AREA

In an attempt to explain the results of this analysis, this section provides a summary of regional and sub-regional sustainability policies and programs in the New York metropolitan area. As already discussed, the entire metro area is under the purview of the Regional Plan Association and includes 31 counties contained within three states (Connecticut, New Jersey, and New York). In addition to this tristate planning agency, the metropolitan areas of Connecticut (the urbanized region in southwest Connecticut) and New York (New York City, the Hudson Valley, and Long Island) developed a New York-Connecticut Sustainable Communities Initiative and received a $3.5 million grant in 2014 from a HUD Sustainable Housing and Communities Regional Plan Grant to develop a regional plan to address sustainable development (New York-Connecticut Sustainable Communities undated). The planning process resulted in a vision for the region, goals and actions for specific place-based projects, and an implementation plan for the bistate metropolitan areas. What follows is a discussion of regional planning and the work of advocacy groups within each of the five RPA subregions.

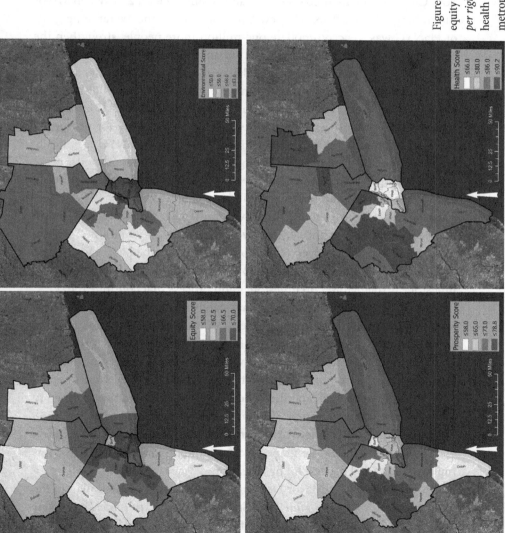

Figure 2.5. Sustainability scores for equity (*upper left*), environment (*upper right*), prosperity (*lower left*), and health (*lower right*) in the New York metropolitan area.

New York City Sub-Region

The five counties/boroughs of New York City are designated as their own sub-region by the RPA and as just discussed participated in the CT-NY Sustainable Communities regional planning effort in 2014. Since the sub-region is entirely under the jurisdiction of city government (as opposed to the many layers of local governments in the other subregions), the highest overall sustainability score (67.5) is not a surprising result. Add to this, New York City has been focused on sustainability since at least 2007 when Mayor Bloomberg released PlaNYC (New York City 2007). The overarching goals of PlaNYC was to accommodate one million more people, strengthen the economy, combat climate change, and enhance the quality of life for all New York City residents. PlaNYC was a collaborative effort with 25 city agencies which was criticized for not including the public (Angotti 2008) and for potentially underrepresenting the Bronx (Rosan 2012) (which is interesting that Bronx County received the lowest sustainability score for prosperity [49.3] of all counties in the metro region). Specifically, PlaNYC specifically targets ten areas of interest, including housing and neighborhoods, parks and public spaces, brownfields, waterways, water supply, transportation, energy, air quality, solid waste, and climate change.

Over 97 percent of the 127 initiatives in PlaNYC were launched within one year of its release and almost two-thirds of its 2009 milestones were achieved or mostly achieved (New York City 2007). The plan was updated in 2011 and was expanded to tackle 132 initiatives and more than 400 specific milestones. The plan set up milestones and process that continue today.

PlaNYC was an aggressive plan that established the framework for long-term sustainability that remained in effect into Mayor De Blasio's mayoral administration until 2015 when the devastation from Superstorm Sandy made it evident that climate resiliency needed a stronger presence in New York City. Thus, in response to Superstorm Sandy, *A Stronger, More Resilient New York* was published, and the Office of Sustainability was established to focus on rebuilding and improve climate resilience (New York City 2013). The program set out to improve infrastructure and resiliency in the medium and long term, as well as enhance local rebuilding and resilience in communities that were hardest hit by the storm.

On Earth Day 2015, Mayor DeBlasio announced an updated plan to the city, *One New York: The Plan for a Strong and Just City (OneNYC)* (New York City 2015b). An aggressive list of the major goals laid out in *OneNYC* is provided below:

Sending zero waste to landfills and reducing waste disposal by 90 percent relative to 2005 levels, by 2030.

Ensuring New York City has the best air quality among all large U.S. cities by 2030.

Reducing risks of flooding in most affected communities.

Eliminating long-term displacement from homes and jobs after future shock events by 2050.

Reducing the city's Social Vulnerability Index for neighborhoods across the city;

Reducing annual economic losses from climate-related events.

Continued investment as part of an over $20 billion program that includes a range of physical, social, and economic resiliency measures.

Making New York City home to 4.9 million jobs by 2040.

Creating 240,000 new housing units by 2025, and an additional 250,000 to 300,000 by 2040.

Enabling the average New Yorker to reach 25 percent more jobs—or 1.8 million jobs—within 45 minutes by public transit.

Lifting 800,000 New Yorkers out of poverty or near poverty by 2025.

Cutting premature mortality by 25 percent by 2040, while reducing racial/ethnic disparities.

Reducing the city's greenhouse gas emissions by 80 percent by 2050, over 2005 levels.

The Mayor's Office of Sustainability has been monitoring progress of its sustainability measures and communicates progress through annual reports. In addition, all data are provided online which adds to government transparency. With planning that focuses on long-term sustainability and strong leadership with effective follow through, New York City is rapidly becoming known as the most sustainable city in the world.

LONG ISLAND SUB-REGION

The Long Island sub-region came in second in overall sustainability (with a score of 66.9) in this analysis. Two regional planning efforts have occurred in this decade. The first plan, *Long Island 2035: Regional Comprehensive Sustainability Plan (LI2035 Plan)* was published in December 2010 (Long Island Regional Planning Council 2010). *LI2035* was a collaborative process that engaged many stakeholders. The plan provides a baseline assessment of the region and a list of strategies organized into the following categories: tax and governance, economy, environment and infrastructure, and equity. The second regional planning effort was complete in May 2013 through a collaborative process. The Cleaner Greener Long Island Regional Sustainability Plan was funded through a New York State Energy Research and Development Authority (NYSERDA) grant (New York State undated a). The plan presents a baseline analysis of Long Island and includes the following focus areas: economic development and workforce housing, energy, transportation, land use and livable communities, waste management, water management, and governance and implementation.

While both these two regional plans are comprehensive and aggressive in its targets, it is unclear who is responsible for implementing the plans which is in stark contrast to New York City which has a dedicated Office of Sustainability to implement the strategies and measures. With over 300 local governments, implementing a regional sustainability plan is challenging; however, there are some local examples of communities that are making progress (e.g., Great Neck Plaza, Huntington, and East Hampton have all adopted plans to address sustainability and climate change).

Two advocacy groups that are important to the advancement of sustainability issues on Long Island include Sustainable Long Island and Vision Long Island. Sustainable Long Island had been serving the area for at least a decade as an educational and networking platform and ran popular annual conferences; however, the organization closed in 2017. Vision Long Island on the other hand has been going strong for close to two decades. This group promotes smart growth development by counseling local governments, educating its stakeholders, and advancing private planning projects across the island. Each year, Vision Long Island hosts a Smart Growth Summit and a Smart Growth Awards ceremony which are very well attended by a diverse array of stakeholders.

HUDSON VALLEY SUB-REGION

The Hudson Valley was slightly less than Long Island with a score of 66.2. In May 2013, the Hudson Valley Regional Council completed a regional sustainability plan, Mid-Hudson Regional Sustainability Plan, through the Cleaner, Greener Communities NYSERDA grant (New York State undated b). The Plan was developed through a collaborative process and provides a baseline analysis and identifies initiatives that include the following categories: land use, livable communities, and transportation; energy; materials management; agriculture and open space; and water. The plan also includes a regional GHG inventory and a climate change vulnerability assessment.

One notable advocacy group in this sub-region is Sustainable Hudson Valley that was formed in the late 1990s to establish a networking platform and provide education to policymakers and the public (Sustainable Hudson Valley undated). The organization highlights a number of programs in the sub-region, including Solarize Hudson Valley, Drive Electric Hudson Valley, Eco-Districts, Blue Economy, and Service Learning. Another advocacy group worth mentioning is Sustainable Westchester which was established in 2016. The group has on its board of directors many of the political leaders in the county and is an indication that the community is supportive of sustainability. This may be one reason that Westchester County ranked second behind Manhattan (New York County) in this analysis.

NORTHERN NEW JERSEY SUB-REGION

The overall score for Northern New Jersey was 65.7 which is second to last. In New Jersey, there is no coordinated statewide strategy specific to sustainability; however, planning and sustainable communities are organizing concepts on the Department of Environmental Protection's web page (New Jersey undated). In terms of regional planning, historically, New Jersey legally delegates land use planning to its counties where some of the counties have developed sustainability plans. The only regional plan prepared specific for the Northern New Jersey region was the Regional Plan for Sustainable Development (RPSD) which was funded by a five-million-dollar grant from the U.S. Department of Housing and Urban Development (HUD) Sustainable Communities Regional Planning Grant Program. The plan used transit-oriented development (TOD) as the framework to promote sustainability in the region (North Jersey Transportation Planning Authority 2015).

New Jersey has three advocacy groups that have been advancing sustainability in the state. PlanSmartNJ, established over 40 years ago, is a planning and research organization that promotes a regional approach to land use. The group currently focuses on economic development, transportation, regional equity, and the environment and advocates for incorporating a clear sustainability vision through comprehensive plans. New Jersey Future is another advocacy group that has been working land use and planning since 1987 (New Jersey Future undated). The group currently focuses on an array of sustainability initiatives that include water infrastructure, rebuilding and resiliency, creating great places to age, revitalizing downtowns, and mainstreaming green infrastructure (New Jersey Future undated).

The most recent sustainability advocacy group, Sustainable Jersey, was established in 2009 (Sustainable Jersey undated). This group provides technical and financial support to local governments (Sustainable Jersey undated). Sustainable Jersey developed a certification program for local governments and to date, they have certified 454 participating communities and 204 certified communities. In addition to the certification program, this group has prepared annual reports on sustainability trends in the state starting in 2015 (Sustainable Jersey undated). The latest report which documented progress toward 57 goals (each measured with indicators) concluded that while there has been some progress, progress was not adequate for 26 of the sustainability goals (Sustainable Jersey 2017). However, this does not mean there are not champion municipal governments that aren't tackling sustainability. For example, the City of Newark issued its sustainability plan in 2013. More municipalities will join Sustainable Jersey with grant funding opportunities. For example, to date over $3.4 million in grants have been awarded to projects throughout New Jersey.

SOUTHWEST CONNECTICUT SUB-REGION

Overall, this sub-region scored the lowest on the sustainability index (64.1). Connecticut as a state does not have a plan dedicated specifically to sustainability; however, they do have the typical state departments that work on sustainability issues (i.e., Department of Energy and Environmental Protection [DEEP] and Department of Community Development). As with other states, there is a state-mandated regional planning process in Connecticut which aggregates towns into regional councils. There are five separate regional planning councils that are either completely or partially contained into RPA's Southwest Connecticut. In other words, the council boundaries in Connecticut do not match the RPA's boundary for Southwest Connecticut. In addition to the state-mandated regional councils, three of the Connecticut councils participated in the CT-NY Sustainable Communities planning efforts as previously discussed.

Since there is no statewide sustainability policy program in the state, a new program, Sustainable CT, was initiated in 2016 to assist all local governments (from small villages to towns) advance sustainability initiatives (Sustainable CT undated). Sustainable CT was created through a collaboration of state government, local governments, and other stakeholders that created a road map of actions, identified financial resources, and developed a local government certification program. The certification program began soliciting applicants in 2018. Sustainble CT's statewide focus and flexibility in serving all types of governments shows promise to advance sustainability throughout the state.

Conclusion

From this balanced sustainability assessment, New York City was determined to be more sustainable than its surrounding suburbs. This finding substantiates through quantitative means the long-held belief that urbanites live more sustainably than their suburban counterparts; however, this finding might be a direct result of the strong sustainability policy in New York City and may not be the case in other metropolitan areas. There is a growing body of literature that suggests that when considering additional criteria such as quality of life that suburbs are closing the gap and may even be more sustainable. For example, Du and Wood's (2017) study in Chicago concluded that high-rise urbanites are not necessarily more sustainable than neighboring suburbanites. This finding indicates suburban life can be more sustainable if specific initiatives are put into place (e.g., produce energy from renewable resources and build multimodal mixed-use communities close to public transportation). From a review of regional plans, the suburbs are embracing smart growth development which shows promise in improving the sustainability in the suburbs.

From the regional policy analysis, New York City is well beyond the planning phase and has been implementing their initiatives for over a decade. Sustainability action in the suburban regions generally is slow and remains stuck in the planning stage. There are close to 1,000 local governments within the New York metropolitan area (Long Island has over 300 alone) who would need to coordinate activities which proves challenging. In addition, these smaller governments often lack human capacity and financial resources to execute the strategies put forth in the regional plans. Thus, the suburbs lack an authoritative body to manage sustainability initiatives. In this void, advocacy groups such as Sustainable Jersey and Sustainable CT are filling this void to organize efforts and provide technical assistance to policymakers. Both these groups have developed a certificate program for municipal governments; however, these programs are voluntary in nature and not all municipalities are participating. But it remains to be seen whether the actions of these groups will result in improved sustainability.

It is important to note that this analysis provides an initial attempt to assess sustainability at the county level at one point in time (namely 2015), and this is the first sustainability index developed for the greater New York City metropolitan area. Further sustainability assessments at even lower geographic scales (i.e., towns, villages, and hamlets) are needed to determine the impacts from each jurisdiction. It would also be helpful to examine how regions have changed through time and whether these planning efforts and advocacy initiatives are effective as well as identify the most effective policy strategies. In addition, additional data would further refine the index presented in this analysis. However, data gaps at the local level continue to plague these types of research endeavors. It is clear though that both the suburbs and the urban core of New York City will continue to grow; therefore, assessing sustainability holistically throughout the region will be critical in ensuring that we meet the challenge of sustainable development into the future.

References Cited

Alexe, Anca. 2016. "Urban Sprawl vs. Compact Cities: What's the Smart Solution?" Last accessed February 22, 2020. http://urbanizehub.com/urban-sprawl-compact-solution/.

Angotti, Tom. 2008. "Is New York's Sustainability Plan Sustainable?" Paper presented to the joint conference of the Association of Collegiate Schools of Planning and Association of European Schools of Planning (ACSP/AESOP), Chicago, July.

Bertelsmann Stiftung and Sustainable Development Solutions Network. Undated. SDG Index and Dashboards Report 2019. Last accessed February 22, 2020. https://www.sdgindex.org/.

Chetty, Raj, Nathaniel Hendren, Patrick Kline, and Emmanual Saez. 2014. "Where Is the Land of Opportunity? The Geography of Intergenerational Mobility in the United States." *Quarterly Journal of Economics* 129(4): 1553–1623.

Du, Peng, and Anthony Wood. 2017. *Downtown High-Rise vs. Suburban Low-Rise Living: A Pilot Study on Urban Sustainability*. CTBUH Research Report. Chicago: Council on Tall Buildings and Urban Habitats.

Glaeser, Edward. 2009. "Green Cities, Brown Suburbs." Last accessed February 22, 2020. https://www.city-journal.org/html/green-cities-brown-suburbs-13143.html.

Green, Jamal. 2018. "Urban vs Suburb: The Debate and Ignoring the Deeper Issues." Last accessed February 22, 2020. https://www.smartcitiesdive.com/ex/sustainablecitiescollective/kotkin-continues-troll-and-we-ignore-deeper-issues/147196/.

Komeily, Ali, and Ravi Srinivasan. 2015. "A Need for Balanced Approach to Neighborhood Sustainability Assessments: A Critical Review and Analysis." *Sustainable Cities and Society* 18: 32–43.

Long Island Regional Planning Council. Undated. Last accessed February 23, 2020. https://lirpc.org/wp-content/uploads/2017/08/LI-2035-Strategies-Report.pdf.

Montemayor, Lucrecia, and Ellis Calvin. 2015. "Identification and Classification of Urban Development Place Types for the New York Metropolitan Area." *EACM Digital Library*: 101–106. https://dl.acm.org/doi/10.1145/2835022.2835039.

New Jersey. Undated. "State of New Jersey Department of Environmental Protection Planning and Sustainable Communities." Last accessed February 23, 2020. https://www.state.nj.us/dep/opsc/index.html.

New Jersey Future. Undated. Last accessed February 23, 2020. https://www.njfuture.org/.

New York City. 2007. *PlaNYC*. Last accessed February 20, 2020. https://www.cakex.org/sites/default/files/documents/full_report_2007.pdf.

New York City. 2013. "A Stronger, More Resilient New York." Last accessed February 20, 2020. https://www.adaptationclearinghouse.org/resources/a-stronger-more-resilient-new-york.html.

New York City. 2014. *Summary of Initiatives with a Complete List of Sustainability Indicators*. Last accessed February 20, 2020. https://www1.nyc.gov/html/onenyc/downloads/pdf/publications/OneNYC-Summary-of-Initiatives.pdf.

New York City. 2015a. *Global Vision, Urban Action: A City with Global Goals*. Last accessed February 20, 2020. https://view.publitas.com/nyc-mayors-office/a-city-with-global-goals-parts-i-and-ii-for-download/page/1.

New York City. 2015b. *OneNYC*. Last accessed February 20, 2020. https://onenyc.cityofnewyork.us/wp-content/uploads/2019/04/OneNYC-Strategic-Plan-2015.pdf.

New York-Connecticut Sustainable Communities Initiative. Undated. Last accessed February 23, 2020. https://www.nymtc.org/Regional-Planning-Activities/Sustainability-Planning/NY-CT-SCI.

New York State. Undated a. "Governor Cuomo Launches Grant Program for Projects to Support Cleaner, Greener Communities in Long Island." Last accessed February 23, 2020. https://www.governor.ny.gov/news/governor-cuomo-launches-grant-program-projects-support-cleaner-greener-communities#.

New York State. Undated b. "Governor Cuomo Launches Grant Program for Projects to Support Cleaner, Greener Communities in Hudson Valley." Last accessed February 23, 2020. https://www.governor.ny.gov/news/governor-cuomo-launches-grant-program-projects-support-cleaner-greener-communities-hudson.

North Jersey Transportation Planning Authority. 2017. "Plan 2045: Connecting North Jersey, NJTPA Regional Transportation Plan." Last accessed February 20, 2020. https://www.njtpa.org/Planning/Plans-Guidance/Plan-2045.aspx.

Organization for Economic Cooperation and Development. 2008. *Handbook on Constructing Composite Indicators: Methodology and User Guide*. Last accessed February 20, 2020. http://www.oecd.org/els/soc/handbookonconstructingcompositeindicators-methodologyanduserguide.htm.

Regional Plan Association. 2017. *Fourth Regional Plan: Making the Region Work for Us*. Last accessed February 22, 2020. http://library.rpa.org/pdf/RPA-The-Fourth-Regional-Plan.pdf.

Rosan, Christina. 2012. "Can PlaNYC Make New York City "Greener and Greater" for Everyone?: Sustainability Planning and the Promise of Environmental Justice." *Local Environment* 17(9): 959–976.

Singh, Rajesh Kumar, H. R. Murty, S. K. Gupta, and A. K. Dikshit. 2009. "An Overview of Sustainability Assessment Methodologies." *Ecological Indicators* 9: 189–219.

Sustainable Development Solutions Network. 2017. "Achieving a Sustainable Urban America: SDSN's U.S. Cities Sustainability Index." Last accessed February 22, 2020. https://resources.unsdsn.org/achieving-a-sustainable-urban-america-sdsns-first-u-s-cities-sdg-index.

Sustainable CT. Undated. Last accessed February 23, 2020. https://sustainablect.org/.

Sustainable Hudson Valley. Undated. Last accessed February 23, 2020. http://sustainhv.org/.

Sustainable Jersey. Undated. Last accessed February 23, 2020. http://www.sustainablejersey.com/.

United Nations. 2018. *Global Indicator Framework for the Sustainable Development Goals and Targets of the 2030 Agenda for Sustainable Development*. Last accessed February 22, 2020. https://unstats.un.org/sdgs/indicators/Global%20Indicator%20Framework%20after%20refinement_Eng.pdf.

United Nations Development Program. Undated. "Sustainable Development Goals." Lasted accessed February 22, 2020. http://www.undp.org/content/undp/en/home/sustainable-development-goals.html.

United Nations Statistics. Undated. SDG Indicators Global Database. https://unstats.un.org/sdgs/indicators/database/.

World Commission on Environment and Development. 1987. *Our Common Future*. Oxford University Press, Oxford.

3

Sustainability through Suburban Growth?

Regional Planning and Targeted Change in the Pacific Northwest

YONN DIERWECHTER

Continued growth in the region can in fact present opportunities for [communities] to restore . . . watersheds, develop more environmentally sensitive approaches to treating stormwater, enhance habitat, and pioneer new technologies and industries that benefit both the environment and the regional economy.

Puget Sound Regional Council (2009a, 10)

The crisis of global unsustainability is upon us. Albeit far too slowly, policies to promote sustainability in turn are shifting at all territorial scales. International green diplomacy, despite well-known reversals and ongoing challenges, has fully emerged (Broadhead 2002) as have national-scale green policy frameworks (e.g., Republic of South Africa 2011). Contemporary global sustainability initiatives now also include city-based networks like C40 and Metropolis that challenge traditional framings of international politics (Amen et al. 2012, Dierwechter 2018, Johnson 2018). "Cities," Herrschel and Newman (2017, ch 1) suggest, "have joined states as international actors." In consequence, traditional "urban" competencies, like land use regulation and public transit services, are increasingly part and parcel of a "global" politics of sustainability (Engelke 2013, Dierwechter 2018).

At the same time, just as global development challenges like sustainability have scaled "down" to involve cities in new ways, some cities have started to scale "up" their otherwise fragmented municipal-scale policies and governance approaches. Leading geographers like Allen Scott (2001) see this as "the rise of city-regions." This thesis signals the importance of managing wider functional economic spaces in ways that produce new spatial coalitions between core cities and their diverse suburbs and rural hinterlands, and thus, novel forms of regional planning and territorial management (Savitch and Adhikari 2017, Herrschel and Dierwechter 2018).

Within this context, the discussion here explores the emerging importance to global sustainability of managing local growth patterns across key metropolitan regions—the real motors of the globalized economy, in Scott's (2001, 813–815) view—through spatially selective regional planning regimes.

In particular, this chapter explores regional planning for more targeted suburban growth across Greater Seattle as a major strategy of sustainability policies. As the head quote above strongly suggests, institutions like the Puget Sound Regional Council in Seattle do not necessarily see growth as a problem *stricto sensu*. If appropriately managed across metropolitan space, they see it as an "opportunity" to benefit both "the environment and the regional economy." This is, of course, a debatable proposition of Smart Growth, an influential regional planning theory of urban development in North America over the last several decades (Fox 2010).

That said, the case study presented below of a low-density, slow-growth community, University Place, demonstrates how *promoting growth in select suburbs* is indeed a crucial component of regional sustainability policy in Greater Seattle. At the same time, that policy paradoxically also means, I shall specifically argue, *reconstructing suburbs* so that they are, in fact, more like cities. The real planning challenge for this key American city-region is not suburbs as political places; it is *suburbia as a material condition*. I explore this main claim in the next section before analyzing regional planning and growth dynamics in the Greater Seattle metropolitan region.

Confronting Suburbia, Wherever Located

This book's attempt to think through the concept and possibilities of sustainable suburbs generates initial questions about the meaning of suburbs as places and/or suburbia as a hypothesized source of unsustainability. As part of my overall themes, I first want to develop the proposition that, in many planning regimes focused on sustainability, the problem is not suburbs; the problem is suburbia, *wherever* located.

Notwithstanding their obvious importance across the world, suburbs as places continue to occupy a thematically ambiguous place in both urban and regional studies. Like sprawl or sustainability, suburbs resist facile treatment. It is difficult to capture parsimoniously their growing empirical variety and functional complexity. It is possible to argue, for instance, that many suburbs have now become cities (or at least more city like). From this perspective, making classic residential suburbs more sustainable is really about slowly turning them into cities (adding new uses). On the other hand, especially in the North American context, a great deal of what constitutes principle (or core) cities is still the low-density, detached, single-family home segregated from the urbanized diversity of nonresidential uses. As Kolko (2015) notes, "Just as big cities contain neighborhoods that feel suburban, some areas outside big-city boundaries—such as Hoboken, New Jersey, and West Hollywood, California—feel more urban than parts of their neighboring big cities." This suggests the real issue—across Greater Seattle at least—is not sustaining suburbs per se but transforming suburbia, *wherever* that condition might be located in the metropolitan arena.

Houston, Texas, the fourth biggest "city" in the United States, is an apt example of how the segregated, low-density architectonics of suburbia, again wherever located, relate to wider concerns with unsustainability and related problems of enervated resiliency. Pummelled by Hurricane Harvey in 2017, the sociopolitical dimensions of this particular natural disaster, notably its freewheeling land use planning policies, received fierce criticism from a number of urban scholars, natural scientists, and local officials (Collier 2017). As a much circulated and nationally discussed analysis by ProPublica and the *Texas News Tribune* claimed:

> Scientists, other experts and federal officials say Houston's explosive growth is largely to blame. As millions have flocked to the metropolitan area in recent decades, local officials have largely snubbed stricter building regulations, allowing developers to pave over crucial acres of prairie land that once absorbed huge amounts of rainwater. That has led to an excess of floodwater during storms that chokes the city's vast bayou network, drainage systems and two huge federally owned reservoirs, endangering many nearby homes. (Sagita et al. 2016)

Seen in these terms, some suburbs around the country are already far more sustainable than cities like Houston. Places like Somerville, Massachusetts, for example, lack the stereotyped features of suburbia commonly identified with unsustainable human settlement patterns: *viz.*, dendritic street patterns, automobile-dependency, severe functional segregation, excessive strip development, low-density monotony, discontinuity, and so on. Following this logic, we are arguably better off if we consider specific forms of suburbia, not suburban jurisdictions. But we should recognize that many (and likely most) jurisdictions in the United States are bereft of the forms and functions commonly associated with the familiar bromides of green land use planning and sustainable cities: mixed use, transit accessibility, compact building envelopes, adaptive reuse, and so on.

Shifting the focus from suburban places to conditions of suburbia generates other interesting insights. According to the U.S. Census Bureau, fifty incorporated municipalities are denser than Boston, which often ranks high on sustainable cities scores; perhaps more surprising, only two of these places (the cities of New York and San Francisco) are traditional cities. The remainder are technically speaking suburbs, but not necessarily *suburban* in the lived density of their physical fabric (and not all these dense suburbs are in the Northeast).

While New York City is obviously much denser than the City of Los Angeles, as a second point, the Los Angeles metropolitan region is in fact denser overall than the New York metropolitan region. The problem, however, is that the Los Angeles (LA) metropolitan area generates what William Fulton has repeatedly called dysfunctional density. Having developed around car technologies, the surprising density of LA is uniform rather than "lumpy" or nodal; Los Angeles lacks a sufficient necklace of nodes dense enough to facilitate, for instance, sufficient region-wide transit alternatives to automobile-oriented development. The New York metropolitan region, in contrast, has twice as many dense places—communities with more than 10,000 people per square mile—than does Los Angeles.

Density is probably a necessary condition of improved metropolitan sustainability (a point that is certainly debatable but I do not explicitly defend here.) Yet increased density without *use* diversity takes space away from people but gives too little in return. Green plans thus focus on enhancing diverse density, which partly relies on channelling new rounds of growth into specifically favored locales.

Shifting the focus from places to conditions also requires shifting attention from local to regional-scale visions of sustainability—and specifically to the role of regional planning in helping to occasion sustainability across local jurisdictions in metropolitan areas. Put in a slightly different way, as twentieth-century cities-with-suburbs built around Fordist regimes of accumulation and regulation have transformed into what Scott (2001, 813) again calls twenty-first-century, post-Fordist global city-regions, the sustainability project has shifted as well. The next section considers this related theme with specific empirical reference to regional planning and growth dynamics across Greater Seattle.

REGIONAL PLANNING IN PUGET SOUND: A NORMATIVE THEORY OF "REGIONAL GEOGRAPHIES"

The liberal political economy in the United States contrasts historically with social democratic and other societal models around the world, such as in Western Europe. That said, city and regional planning regimes do differ within the American polity, as land management approaches are shaped by variegated state-level legal and policy dynamics (Ingram et al. 2009).

In their review of the largest 50 metropolitan regions in the United States, for example, Pendall and Puentes (2008) present diverse "families" of land use regimes as distinctive forms of territorial regulation. They include in their "reform family" major metropolitan areas like Greater Seattle, known officially as the Puget Sound region after the inland sea around which the urban area has developed. Like metropolitan Portland to its south, the Puget Sound region as a whole is today committed to the "containment" of exurban sprawl through regulatory tools like urban growth boundaries (other "reform" cases, such as Baltimore, are characterized only by "contain-lite" powers while most metropolitan areas in the United States deploy either "traditional" or indeed "exclusionary" planning regimes). Urban growth boundaries and similar planning techniques (such as urban service lines) are deployed in many individual communities, including suburbs; however, they are rarely coordinated de jure across entire metropolitan regions (Dierwechter 2008). Efforts to "green up" suburban growth in the Pacific Northwest are therefore tied closely to regional planning programs that have arguably strengthened institutionally since landmark passage of the Growth Management Act in 1990/1991.

Washington's Growth Management Act (GMA) has been crucial. It mandates comprehensive planning in most counties and cities, including countywide planning policies; regionally coordinated urban growth boundaries through county-scale administrations; service concurrency provisions; and the protection of critical areas, particularly shorelines sheltered earlier under the separate but related Shoreline Management Act. This regulatory system, which drew heavily on the planning experiences of Florida and Oregon, today represents an important milestone in the institutional and legal history of American planning and indeed pragmatic environmental thought and action.

However, the larger GMA system, as it is often called locally, also reflects a wider ideological shift in the United States away from the more state-centric, top-down, neo-Keynesian model of planning built in neighboring

Oregon in the early 1970s, often widely considered the exemplar of urban sustainability policies through planning visions, especially in Portland (Ozawa 2004, Nelson and Moore 1992). Built in the 1990s rather than 1970s, Washington's planning system reflects a stronger concern with protecting property rights, greater emphasis on bottom-up coordination, and the absence of sufficiently toothy provisions for the creation of authoritative regional governments with direct oversight powers in policy matters of metropolitan-scale planning and community development priorities.

So, does this planning system advance sustainability "at the sub-national level" (Eran 2017)? In theory, some planning scholars in Washington have suggested that it does—or at least that it might do so over time (Miller 2004). In practice, the record is more equivocal. Empirical work on the sociospatial and ecological impacts of Washington's land use planning initiatives over the past 25 years of actual policy implementation on the ground reflects the political compromises, institutional tensions, and perhaps even historic rigidities inherent structurally within the wider GMA system.

On the one hand, for example, scholars have documented significant early reductions in development permits issued *outside of* growth boundaries in key urban counties of the state (Carlson and Dierwechter 2007). Other researchers, however, note that while housing density has increased within these boundaries across Greater Seattle, "policies to reduce the density of settlement outside urban centers, in part to protect ecological systems, may have unintended environmental consequences" (Robinson, Newell, and Marzluff 2005, 76). My own studies of regional growth policies refer to "smart segregation" within Seattle even as I also document more balanced experiences in other communities (Dierwechter 2017, 2014).

For their part, many local practitioners see real possibilities in building "equitable transit communities" (Bakkente et al. 2015). Yet Seattle's dramatic economic transformation into one of the country's leading high-tech hubs has generated considerable anxiety that planning policies focused originally on urban regeneration are today ill equipped to ensure social equity and inclusive growth (Fowler 2015, Abel, White, and Clauson 2015, Balk 2014, Klingle 2007). Rising homelessness, uneven climate action, infrastructure decay, and the slow rollout of public transit alternatives have generated recursive political and policy concerns, especially when linked with wider sustainability discourses.

These broader themes are important when considering specific regional planning efforts. The area's federally designated Metropolitan Planning Organization, the aforementioned Puget Sound Regional Council (PSRC), along with key service institutions like Sound Transit, a regional transit agency with voter-strengthened taxing authority, highlight the crucial importance of supralocal public authorities in the "piecemeal, partial, and selective" reshaping of otherwise uneven growth dynamics in complex metropolitan areas (Savitch and Adhikari 2017, 348–349). Public authorities like the PSRC carry *within* them the institutional and territorial logic of regional (or metropolitan-scale) planning itself, even as local planning powers remain intact. Savitch and Adhikari consider such a regional planning logic an "unbundled function" that necessarily pushes up and away from "local autonomy" in order to focus on the larger structural need for "better coordination, common policies, and shared public goods" (Savitch and Adhikari 2017).

That makes for inevitable friction. The tension between the (social and cultural) desire for local autonomy, which can result in subtle exclusion or even outright extrusion, and the (economic and ecological) demand for regional planning produces, in their terminology, a fragmented rather than integrated regionalism.

Elsewhere I have argued, however, that planning policy attempts to deepen sustainability across Greater Seattle, as in other U.S. city-regions, are perhaps more usefully explained as the product of the "intercurrence" of institutions (Dierwechter 2017), a term derived from the work of Karen Orren and Steve Skowronek (1996, 2004). Intercurrence refers to "multiple orders in simultaneous action," that is, a world of "ordered disorder," where relatively independent institutions move in and out of alignment with one another in patterns of both continuity and change (Rogers 2005). The politics of regional planning for sustainability, then, are the politics of intercurrence, wherein both ideational and institutional efforts to overcome the "abutting and grating" of institutional authority are always at work (Orren and Skowronek 1996).

A regional planning example to try to manage (if never resolve) the frictions of intercurrence across Greater Seattle is the PSRC's spatial concept and investment program of so-called Regional Geographies. Informed by architectonic and urban form concepts like Smart Growth, New Urbanism, and urban sustainability principles, as well as procedural ideals like rational comprehensive decision-making and collaborative planning, PSRC's attempt to reshape the form and function of Greater Seattle illustrates how planners and other governance actors envision suburban sustainability at multiple spatial and regulatory scales.

The PSRC lacks direct land use authority. But the PSRC's regional plan for urban growth management, VISON 2040 (Puget Sound Regional Council 2009a), along with the GMA and consistently focused federal transit support (Bakkente 2012), all emphasize the concentration of new housing, jobs and services within the region's urban growth boundaries and especially within key growth centers. This synoptic development goal also aims to deliver infrastructure more efficiently. The self-stated "hallmark" of this strategy, moreover, is the negotiated distribution of growth around Regional Geographies, which is defined by the simple idea that "different types of cities and unincorporated areas will play distinct roles in the region's future" (Puget Sound Regional Council 2009a, 17). So theoretically imagined, the plan claims, "continued growth in the region can in fact present opportunities for [communities] to restore . . . watersheds, develop more environmentally sensitive approaches to treating stormwater, enhance habitat, and pioneer new technologies and industries that benefit both the environment and the regional economy" (Puget Sound Regional Council 2009a).

The location of growth is thus everything. As new urban growth is shunted *away* from more peripheral areas in the region through multicounty urban growth boundaries and other tools, it is also directed *toward* favored centers of various kinds within select communities that, over time, are better integrated with transit alternatives—including commuter, bus rapid transit, and light rail systems. This is illustrated simplistically in Figure 3.1 below. What should be immediately clear from this vision is that while some suburbs are expected to become more sustainable *without* significant growth and attendant progrowth policies, others are expected to become sustainable *through* growth—the main theme of this chapter. This occurs by reshaping, redesigning and indeed better integrating such growth in ways that comport with recent planning theories for sustainability like Smart Growth.

One such community is the traditional suburban municipality of University Place, located south of Tacoma in a part of the metropolitan area, Pierce County, that heretofore has arguably grown relatively too slow to comport with the Regional Geographies theory of a more sustainable metropolitan area. As the next section explicitly shows, "amping up" long-dormant local growth rates in University Place through coordinated local and regional planning policies is now a crucial part of how Greater Seattle imagines and implements sustainability.

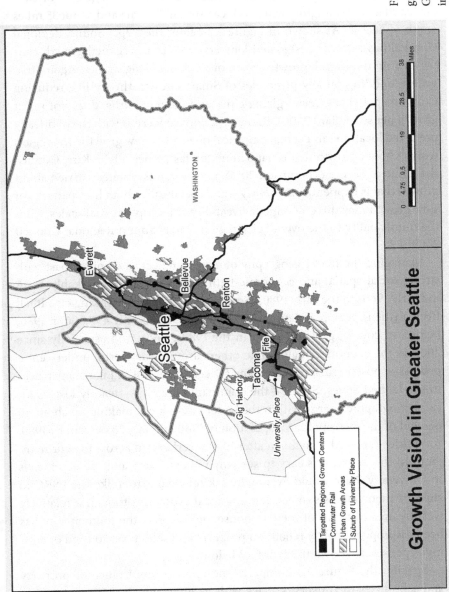

Growth Vision in Greater Seattle

Figure 3.1. Sustainable growth vision across Greater Seattle, Washington State.

Suburban Regional Growth in University Place, Washington

The idea that select suburbs should more aggressively promote new growth through a closely negotiated regional planning regime *in order to* advance the global project of sustainability is illustrated in the specific case of University Place, Washington, which is located near Tacoma and about 35 miles south of Seattle. As shown in Figure 3.2 below, the Puget Sound Regional Council has recently designated University Place a "regional growth center" within the overall growth vision of Regional Geographies. Again, that vision, building on key principles of Smart Growth theory like reducing exurban sprawl and reenergizing cities and older suburbs, does not reject growth per se (Pollard 2000). Rather, the core concern is with the politically preferred location (and actual empirical quality) of new growth. The region worries about growth that is happening too fast, especially in King County, the home of Seattle (Thompson 2016); however, it is equally worried about growth that is happening *too slowly*—and specifically about how patterns of slow growth may directly impede liveability at multiple spatial scales, from the municipality to the overall city-region (Puget Sound Regional Council 2009b).

Managing the future geography of growth reflects a complex past with strong social-spatial and economic-path dependencies. Remarkably, from the early 1960s to the early 1980s, Seattle was actually a shrinking city. It fell from a (then) peak population of 557,000 in 1960 to only 497,000 in 1980, before, of course, growing again in the 1980s, 1990s and especially since 2000 as the economic region, like others around the world, underwent a process of post-Fordist restructuring (Dierwechter 2017). Like many places in the United States, Seattle—the dominant core city—steadily lost its local demographic energy within the wider city-region, making up about 40 percent of the metropolitan population in 1960 but only 23 percent by 1980.

Seattle's renewed dynamism after 1980 reverberated across the entire region. In this sense, it is easy to see why policymakers and public officials in University Place would eventually ask regional actors like the PSRC to support their interests in accelerating local growth. Although a relatively wealthy area in terms of median household income, the municipality has been disproportionately reliant on property and sales taxes to fund operating expenses, as shown in Figure 3.3 below.

Specifically, Figure 3.3 compares the relative importance of property and sales taxes in University Place with neighboring Gig Harbor, another wealthy suburb in the area. Paradoxically, as shown separately in Figure 3.4,

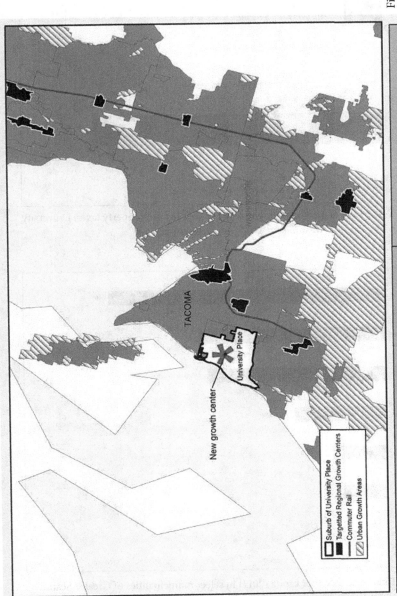

Figure 3.2. The slow-growth suburb of University Place in regional policy context.

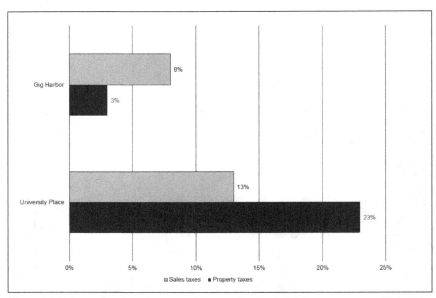

Figure 3.3. Percentage of total operating revenue from sales and property taxes: University Place versus Gig Harbor, Washington, 2017 (calculated by author).

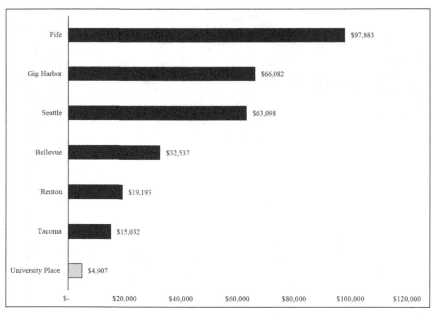

Figure 3.4. Total retail sales per capita (2012) in select municipalities of Greater Seattle area. (Source: https://www.census.gov/quickfacts/fact/table/seattlecitywashington,US/RTN130212.)

University Place's retail base is also still quite small relative to both adjacent suburbs like Gig Harbor, Fife, and Renton as well as bigger cities like Seattle and Tacoma. This weak retail base reflects University Place's origins as a potpourri of postwar residential cul-de-sacs permitted incrementally in "unincorporated" Pierce County until the mid-1990s, when the area, in the wake of the GMA, voted to incorporate in order to take more direct control over its land use, environmental, and economic development planning.

University Place was never really a bounded "community" with strongly articulated elements of residential, commercial, political, and cultural life. Over the past 25 years or so its policy efforts to mature into a more coherent space have been stymied significantly by older path-dependencies and patterns—an example of how problems of intercurrence work. Although most communities in Washington incorporate "economic development" as a key policy element within their overall comprehensive plan and growth management system, University Place established an Office of Economic Development to oversee strategic efforts to reverse this developmental history. In 2007, the second of four major planning efforts by this office identified the still-new municipality's overriding challenge:

> The City's current economy is small relative to its population base, with the majority of UP employees working outside of the City. There is also significant sales tax leakage occurring as City residents do much of their shopping outside City limits. While known for being an affluent community, the City is also home to a surprisingly large population that lives on very moderate incomes. Housing, employment, and shopping opportunities for this population are important to maintaining a diverse and self-sufficient community. (City of University Place 2007, 8)

The 2007 strategic plan for economic development was standard in many respects. It included common goals that sought to encourage a more diverse mix of businesses (6), more concerted efforts to promote a business climate supportive of economic growth (10), and the now always-present "branding" exercises of local economic development activities (18). Such policies comport with the massive literature on the role of enhancing competitiveness in local urban policies and growth planning policy patterns (York, Feiock, and Steinacker 2013). More interesting, then, is the plan's attention to the municipality's role and responsibilities within a wider regional economy and growth management framework. In other words, local policy conversations about planning and development started to reflect the institutional opportunities afforded by—and ideational effects associated with—the PSRC's overall vision of a more sustainable region.

Policy emphasis now focused on meeting "minimum thresholds" to constitute a more densely developed and walkable growth center. Officials in University Place specifically sought over several years to meet a minimum threshold of 18 activity units (jobs plus population) per acre. They also demonstrated to others through well-developed plans and political commitments exactly how they would actually achieve over time at least 45 activity units per acre. At the time of its application, University Place only had about 0.53 jobs for every resident compared with an overall average of 2.67 jobs per resident for all regional growth centers across Greater Seattle. It therefore had to promise, in essence, to stop being a traditional suburb with few jobs and instead focus its main policy efforts on growing both people and jobs. It had to promise to become more city-like (Puget Sound Regional Council 2014, 7). It did just that, year after year.

These efforts bore fruit. In 2014, the PSRC agreed to designate a 465-acre mixed-use area within University Place as a "Provisional Regional Growth Center," a major planning policy decision for both the municipality and the region that literally put this suburb on the strategic "growth" map. "Provisional" status is ultimately removed when municipalities selected for concentrated growth support from federally supported regional authorities like the PSRC develop and formally adopt a rigorous subarea plan within two years. Granted an extension, University Place subsequently published that plan in late 2017. The plan targeted relatively aggressive goals in three key nodes within the boundaries of the footprint of the new growth center.

In particular, the PSRC and University Place now conceived growth as more "activity units" per acre, measured as most any synergistic complementarity between residential housing units and new jobs. The approval of the growth application meant that University Place, despite its relatively small and slow-changing economy, would now receive preferential future treatment (within its own county and across the entire metropolitan region), and especially in the annual dispersal of federal transit infrastructure investments. This is a key point. The approval now meant that the wider region looked upon a once ill-defined patch of suburbia as a future city.

Regional politics not discussed here were and remain important. Yet years of deliberated local planning and development priorities to rectify the city's tax leakage problem are also significant in this multiscalar policy story. Formally, the city amended its comprehensive plan in 2010 to visualize a future growth center. This process included (and will continue to include) remaking many suburban development policy and regulatory code inconsistencies (e.g., toward new design standards for pedestrian-friendly, transit-supportive developments). Perhaps more importantly, the city's capital improvement disbursements have included support for high-occupancy vehicle use, local transit, regional high-capacity transit, and nonmotorized transportation—all *urban* policy initiatives heavily consistent with contemporary discussions of *urban* sustainability.

In justifying its decision to designate University Place as a new regional growth center, the PSRC pointed logically to its consistency with a regional planning vision that champions "compact communities" which are "relatively densely developed and walkable . . . [where] most future residential and employment growth will occur" (Puget Sound Regional Council 2014). The PSRC supports this suburb, in other words, because it has committed itself to a postsuburban policy agenda commensurate with current theories of sustainable cities.

CONCLUSIONS

What is the role of the American suburb in the broader search for a more sustainable global order? In dynamic global city-regions like Greater Seattle, answering this question depends on how fast or slow a particular suburb is growing demographically and economically; on where exactly that suburb is located within the wider and uneven regional space-economy; and, finally, on whether or not that particular suburb can convincingly plan for new "postsuburbia" elements of greener development, notably mixed-use employment and residential areas directly supportive of greater economic self-sufficiency through improved jobs-housing balance as well as enhanced transit accessibility and local walkability. While untrammelled, uncoordinated, and poorly deliberated progrowth policies, especially in American suburbs, have been a major cause of both ecological decay and socioeconomic segregation (Dierwechter 2014), continued growth of key suburbs is also paradoxically mobilized as a key strategy to reshape the inherited spatialities of select city-regions into more sustainable forms.

Suburbs like University Place in Pierce County, Washington, are now targeted for more aggressive growth trajectories by multiscalar planning regimes. These regimes seek to jettison much of the suburban conditions that have traditionally hamstrung their own economic health and fiscal position as well as the overall functionality of the wider city-region. Embedded within a putatively smart regional planning regime committed to a normative development concept of Regional Geographies, wherein different kinds of communities perform different kinds of coordinated and interrelated roles, sustainability occurs, in part, *through new suburban growth*—perhaps an odd-sounding statement when stripped of its regional context. Such growth, however, is not simply more of the same; it is a consciously deliberated mix of uses and functions that asks some suburbs to become, in my judgement, more "urban" even as they seek to balance these demands with local expectations and cultural mores that will struggle significantly to embrace these new demands and policy realities.

This last tension will shape the difficult politics of future suburban change. "We see all that heated growth up there in Seattle," the city manager of University Place remarked publicly in early 2016, "and we all want to see if we can get more of that growth down here in our community, but also protecting our single family residential areas" (Sugg 2016). The structural tension of selectively changing within the context of wider constancy; of targeting new rounds of economic and residential growth without unleashing a torrent of new anti-growth politics—all this again suggests the uneasy comingling of multiple political orders, or what Orren and Skowronek (2004) call the intercurrence of divergent institutional norms that strongly characterize American political developments at all spatial scales. Confronting the crisis of global unsustainability by confronting the spatialities of local suburbia through coordinated regional planning programs, in Seattle as elsewhere, may fall well short of the radical changes needed. But, it will be hard enough.

References Cited

Abel, Troy, Jonah White, and Stacey Clauson. 2015. "Risky Business: Sustainability and Industrial Land Use Across Seattle's Gentrifying Riskscape." *Sustainability* 7(11): 15718–15753.

Amen, Mark, Noel Toly, Patricia McCarney, and Klaus Segbers, eds. 2012. *Cities and Global Governance: New Sites for International Relations*. Farnham, UK: Ashgate.

Bakkente, Ben. 2012. Senior regional planner, PSRC. Seattle. Interview. April 23.

Bakkente, Ben, Mary Pat Lawlor, Amanda Roberts, and Miranda Redinger. 2015. "Growing Transit Communities." In *Planning the Pacific Northwest*, edited by J. Sterrett, C. Ozawa, D. Ryan, E. Seltzer, and J. Whittington. Chicago: APA Press.

Balk, Gene. 2014. "As Seattle Gets Richer, the City's Black Households Get Poorer." *Seattle Times*, November 12. https://www.seattletimes.com/news/as-seattle-gets-richer-the-citys-black-households-get-poorer-2/.

Broadhead, Lee-Anne. 2002. *International Environmental Politics: The Limits to Green Diplomacy*. Boulder, CO: Lynne Rienner.

Carlson, Tom, and Yonn Dierwechter. 2007. "Effects of Urban Growth Boundaries on Residential Development in Pierce County, Washington." *Professional Geographer* 59(2): 209–220.

City of University Place. 2007. Economic Development Strategic Action Plan. Edited by Economic Development Office. University Place, WA.

Collier, Kiah. 2017. "When Climate Change Meets Sprawl: Why Houston's 'Once-in-a-Lifetime' Floods Keep Happening." GreenSourceDFW. Last accessed October 22, 2017. https://www.greensourcedfw.org/articles/when-climate-change-meets-sprawl-why-houstons-once-lifetime-floods-keep-happening.

Dierwechter, Yonn. 2008. *Urban Growth Management and Its Discontents: Promises, Practices and Geopolitics in US City-Regions.* New York: Palgrave.

Dierwechter, Yonn. 2014. "The Spaces That Smart Growth Makes: Sustainability, Segregation, and Residential Change Across Greater Seattle." *Urban Geography* 35(5): 691–714. DOI: 10.1080/02723638.2014.916905.

Dierwechter, Yonn. 2017. *Urban Sustainability Through Smart Growth: Intercurrence, Planning, and the Geographies of Regional Development Across Greater Seattle.* Cham, Switz.: Springer.

Dierwechter, Yonn. 2018. *The Urbanization of Green Internationalism.* New York: Palgrave.

Engelke, Peter. 2013. *Foreign Policy for an Urban World: Global Governance and the Rise of Cities.* Washington, D.C.: Atlantic Council.

Eran, Feitelson. 2017. *Advancing Sustainability at the Sub-National Level: The Potential and Limitations of Planning.* Aldershot, UK: Taylor and Francis.

Fowler, Christopher. 2015. "Segregation as a Multiscalar Phenomenon and Its Implications for Neighborhood-Scale Research: The Case of South Seattle, 1990–2010." *Urban Geography.* Published online before press. DOI: 10.1080/02723638.2015.1043775.

Fox, David. 2010. "Halting Sprawl: Smart Growth in Vancouver and Seattle." *Boston College International and Comparative Law Review* 33(1): 43–59.

Herrschel, Tassilo, and Yonn Dierwechter. 2018. *Smart Transitions in City-Regionalism: The Quest for Competitiveness and Sustainability.* London: Routledge.

Herrschel, Tassilo, and Peter Newman. 2017. *Cities as International Actors.* London: Palgrave.

Ingram, Gregory, Armando Carbonell, Yu-Hung Hong, and Anthony Flint. 2009. *Smart Growth Policies: An Evaluation of Program Outcomes.* Cambridge, MA: Lincoln Institute of Land Policy.

Johnson, Craig. 2018. *The Power of Cities in Global Climate Politics.* New York: Palgrave.

Klingle, Matthew. 2007. *Emerald City: An Environmental History of Seattle.* New Haven: Yale University Press.

Kolko, Jed. 2015. "How Suburban Are Big American Cities?" *Huffington Post.* Last accessed May 22, 2016. https://fivethirtyeight.com/features/how-suburban-are-big-american-cities/.

Miller, Donald. 2004. "Decentralized Planning for Sustainable Urban Development: The Washington State Growth Management Act and Its Application in the Seattle Metropolitan Area." In *Advancing Sustainability at the Sub-National Level: The Potential and Limitations of Planning,* edited by Eran Feitelson. Aldershot, UK: Taylor and Francis.

Nelson, Arthur, and Terry Moore. 1992. "Assessing Urban Growth Management: The Case of Portland, Oregon, the USA's Largest Urban Growth Boundary." *Land Use Policy* October: 293–302.

Orren, Karen, and Steve Skowronek. 1996. "Institutions and Intercurrence: Theory Building in the Fullness of Time." *Nomos XXXVII, Political Order*: 111–146.

Orren, Karen, and Steve Skowronek. 2004. *The Search for American Political Development.* Cambridge: Cambridge University Press.

Ozawa, Connie, ed. 2004. *The Portland Edge.* Washington, D.C.: Island Press.

Pendall, Rolf, and Robert Puentes. 2008. "Land-Use Regulations as Territorial Governance in US Metropolitan Areas." *Boletín de la A.G.E.N.* 46: 181–206.

Pollard, Oliver. 2000. "Smart Growth: The Promise, Politics, and Potential Pitfalls of Emerging Growth Management Strategies." *Virginia Environmental Law Journal* 19: 247–285.

Puget Sound Regional Council. 2009a. *VISION 2040.* Seattle: Puget Sound Regional Council.

Puget Sound Regional Council. 2009b. *VISION 2040 and Centers.* Seattle: Puget Sound Regional Council.

Puget Sound Regional Council. 2014. *Regional Designation Recommendation Report: Review of Regional Growth Center Application.* Edited by PSRC. Seattle: PSRC.

Republic of South Africa. 2011. *National Strategy for Sustainable Development and Action Plan.* Pretoria: Department of Environmental Affairs.

Robinson, Lin, Joshua P. Newell, and John M. Marzluff. 2005. "Twenty-Five Years of Sprawl in the Seattle Region: Growth Management Responses and Implications for Conservation." *Landscape and Urban Planning* 71(1): 51–72.

Rogers, Daniel. 2005. "The Search for American Political Development (review)." *Journal of Interdisciplinary History* 36(2): 275–276.

Satija, Neema, Kiah Reveal, Al Shaw, and Jeff Larson. 2016. "Hell and High Water." https://projects.propublica.org/houston/.

Savitch, Hank, and Sarin Adhikari. 2017. "Fragmented Regionalism." *Journal of Urban Affairs* 53(2). DOI: 10.1177/1078087416630626.

Scott, Allen. 2001. "Globalization and the Rise of City-Regions." *European Planning Studies* 9(7): 813–826.

Sugg, Steve. 2016. Public speech. University Place, Washington, April 16.

Thompson, Lynn. 2016. "Regional Council Warns 5 King County Towns They're Growing Too Fast." *Seattle Times*, June 22. https://www.seattletimes.com/seattle-news/eastside/regional-council-warns-5-king-county-towns-theyre-growing-too-fast/.

York, Abigail M., Richard C. Feiock, and Annette Steinacker. 2013. "Dimensions of Economic Development and Growth Management Policy Choices." *State and Local Government Review* 45(2): 86–97. DOI: 10.1177/0160323X13479956.

4

Linking Smart Growth Policies and Natural Disasters

A Study of Local Governments in Florida

VASWATI CHATTERJEE, SIMON A. ANDREW, AND RICHARD C. FEIOCK

Climate change and disaster risks are closely interlinked phenomena. Solecki, Leichenko, O'Brien (2011), for example, identify increased probability of occurrence of disasters as manifestations of climate change (136). Changing weather conditions have increased incidence of extreme events that have interacted with population vulnerabilities and thus resulted in disasters. According to Schipper and Pelling (2006, 20–21), "Successful mitigation of anthropogenic climate change can decrease disaster risk directly by reducing weather-related uncertainty and hazard, and by diminishing the threat of asset depletion among vulnerable natural resource dependent societies; indirect influence comes from the impacts of climate change on national development and, consequently, the asset base available for building resilience and coping with disasters."

With the formulation and implementation of the Kyoto Protocol in 2005 and the *Hyogo Framework for Action 2005–2015*, there has been increasing focus internationally on climate change and disaster risk reduction. While the Kyoto framework highlighted the importance of interorganizational efforts to tackle challenges posed by climate change, the Hyogo framework promoted strategic and systematic efforts to reduce vulnerabilities toward risk to hazards. With the recent United Nations General Assembly endorsement of the *Sendai Framework for Disaster Risk Reduction 2015–2030*, much of the global efforts are turning to regional governance structures for climate change adaptation (CCA) and disaster risk reduction (DRR) policies. With the emergence of regional governance roles in climate change policies, smart growth policies have received increasing focus (Outka and Feiock 2012).

However, very little has been done to understand the association between climate change, smart growth policies, and risk reduction strategies at the local level. The emphasis of smart growth policies to mitigate the effect of climate change is not new. Smart growth policies—broadly defined by a set of 10 principles—mostly focus on policies related to green infrastructure investment, urban development, technologies, and a reduction in traffic congestion and building livable communities (American Planning Association 2012). Scholars have pointed out that smart growth policies believe to be able to reduce greenhouse gas (GHG) emissions and insured losses from natural disasters (Brown and Southworth 2008; Burby 2005; Coaffee 2008). Other forms of smart growth policies contributing to climate change policies include a reduction of vehicle miles travelled, compact development, and optimizing infrastructure requirement for development.

The main argument of this chapter is that extreme weather–related events influence the decision of local governments to adopt certain types of smart growth policies (e.g., mitigation and adaptation). The assumption is that smart growth policies, when translated into mitigation and adaptation approaches, are not homogenous but generate tangible or intangible benefits to certain interest groups. While the literature on disaster and urban management has been employed to explain the incentives and motivation of local governments adopting smart growth policies, we extend the argument by arguing that disasters create a window of opportunity for policy change.

This chapter is guided by the following questions: What smart growth policies have cities undertaken in response to extreme climate related weather events? Are there associations between smart growth policies and natural disasters? These questions not only provide an overview of policy options associated with weather-related events, they also highlight the motivation and incentives of multiple interest groups in coordinating policy preferences. By answering these research questions, policymakers can be better informed on how to translate the concept of smart growth policies in the context of climate change.

The empirical analysis is based on data collected among 165 municipalities in Florida. The results based on nonparametric analysis suggest that climate change mitigation policies appear to have an association with incidence of natural disasters, especially in a policy area that is related to infill development. We also found that local governments that experienced natural disasters but did not require/encourage GHG emission policies tend to have a lower amount of property damages compared to those that have policies related to reduction in GHG emission. Similar results on property damages can be found among local governments with compact development policy and LEED/Green certification for new development/redevelopment policy. As for the association between climate change adaptation policy and natural disasters, we found that a significant difference in property damages between local governments with mechanisms to encourage green construction and technology related to reduced water use and permeable paving.

The next section provides an overview of climate change, smart growth policies, and natural disasters. The chapter then explains the hypothesis related to the association between natural disaster and climate change adaptation and mitigation policies. Before the conclusion section, we present the research design and analysis.

Climate Change, Smart Growth Policies, and Natural Disasters

Local jurisdictions and their constituents have drawn much of the attention to policies related to smart growth as it relates to climate change policy as well as disaster risk reduction activities. According to Schipper and Pelling (2006, 20), "Disaster risk reduction is largely a task for local actors, albeit with support from national and international organisations, particularly in humanitarian action." Outka and Feiock (2012) point out that states in the U.S. have delegated responsibilities in land-use planning (the major tool for climate change policies) to local jurisdictions. Local governments have jurisdictional power to influence individuals and corporate behaviors in their actions toward adopting sustainable development because of their proximity to local constituents. They also enforce building codes that directly affect GHG emissions.

However, smart growth policies involve multiple instruments. The American Planning Association, for instance, identifies that smart growth involves multiple objectives that revolve around efficient and sustainable land development, infrastructure investments, and preservation of open spaces. Each of these objectives are catered through policies in multiple sectors like housing, transportation, community development, and so forth. At the same time, implementing a broad set of smart growth policies—in order to reduce the consequence of disasters—has political and adverse policy consequences.

Burby et al. (2001) suggest that, if not carefully considered, certain smart growth policies like urban containment can increase exposure of communities to hazards. Since different tools of smart growth align themselves with different categories of climate change policies, they are often contradictory to each other. Political consequences of smart growth adoption often involve balancing conflicting priorities of multiple interest groups involved in the process. Proponents of smart group policies include interest groups[1] advocating for land preservation and protection of green spaces, including environmental protection groups, college graduates, and neighborhood associations. Political opposition to smart growth is mostly associated with restriction to urban sprawl from business development and property rights groups (O'Connell 2009; Park, Park, and Lee 2012; Ramirez de la Cruz 2009).

However, few scholars have linked smart growth policies to natural disasters at the local level. Disaster scholars, for example, identify extreme events as "windows of opportunity" wherein policies are adopted after extreme events due to sudden policy focus in this area. It can be argued that, since hazards are manifestations of changing climate (O'Brien et al. 2006), there is a possibility that local jurisdictions will integrate climate change policies with disaster risk reduction management plans. Moreover, O'Brien et al. (2006) and Schipper and Pelling (2006) argued that climate change policies should be integrated with disaster management in order to enhance community resiliency and mitigate risk. Berke, Kartez, and Wenger (1993) suggest that disaster incidences lead to better recovery investment, zoning strategy, and adaptation of land-use policies to "fit" postdisaster requirements.

Another argument is that local governments' smart growth policies share similar goals with hazard mitigation. While smart growth aims at providing better living condition for communities, hazard mitigation contributes to the same objective, that is, lessening a community's exposure to risk from changing climatic conditions. Both the policies utilize tools spanning multiple sectors like transportation, land use, construction, and so forth. Some scholars have also argued that land-use planning is an important tool in order to build stronger communities toward disaster risks (Burby et al. 2000). Since smart growth policies are comprehensive tools that involve sustainable land-use development and improvement in quality of life for communities, they help in advancing hazard mitigation initiatives after disasters.

However, different smart growth tools contribute to hazard mitigation differently. Conflicting strategies can also emerge. While hazard mitigation policies support compact development to reduce urban sprawl, other forms of smart growth tools like introducing heat islands (open areas) within the urban area may work against compact development. Hence, a proper analysis of smart growth tools that work toward mitigation should be considered in order to avoid unintended consequences (Hamin and Gurran 2009). Because of increased focus and new opportunity to implement mitigation policies after disasters, it is expected that there will be an increased adoption of smart growth tools that contribute toward mitigation after disasters.

CLIMATE CHANGE MITIGATION AND ADAPTATION POLICIES

In the context of smart growth planning, what are the differences between climate change mitigation and adaptation policies? Scholars have categorized local response to climate change into two major concepts: climate change adaptation and climate change mitigation (Larsen and Ostling 2009; de Oliveira 2009; Hamin and Gurran 2009; Laukkonen et al. 2009). For instance, climate change mitigation refers to reducing factors contributing to climate change, like greenhouse gas emissions, through control of vehicle transmissions and regulating the built environment. Climate change adaptation, on the other hand, refers to adjusting built and social environment to minimize negative impacts of unavoidable climate change effects (implying changes that are already happening). Adaptation policies may include tools like counteracting effects for rising water levels in coastal areas and creating larger open spaces within the urban fabric for environmental cooling effects in response to rising global, that is, temperatures (Hamim and Gurran 2009).

Mitigation policies with their goal of reducing GHG emissions require compact urban forms of development that help to reduce vehicle transit. Climate change adaptation policies, on the other hand—with the goal of controlling rising heat due to global warming—often focus on the implementation of open spaces within the urban fabric to reduce heat island impacts. In other words, while mitigation policies might advocate compact urban forms, adaptation policies promote a dispersed development pattern thus causing both categories of policies to conflict with each other.

However, climate change mitigation and adaptation policies are often associated with externalities, that is, horizontal, vertical, and functional externalities (Feiock 2013). For example, horizontal externalities can be linked to policy tools such as a reduction of GHG and vehicular emissions. They can reduce pollution and improve regional climate. They can also lead to free-rider problems from other jurisdictions. Similarly, policies related to pedestrian walkways and public transit systems may improve livability of surrounding communities and foster regional economies (Frank 2000; Shapiro 2006). Other jurisdictions may contribute very little to the implementation of such policies and still reap the benefits (Barbour and Teitz 2001). Similarly, compact development may incur negative externalities like reduced regional housing affordability (Dawkins and Nelson 2002).

Climate change mitigation and adaptation policies can also lead to vertical externalities, particularly when smart growth policies are pursued by state, regional, and local governments together. In Florida, for example, state and regional-level agencies are strong advocates of smart growth policies, suggesting that local governments will create positive vertical externalities if they also propagate such policies. At the same time, vertical externalities occur because local governments will be motivated to implement climate change policies when higher level governments (federal/state/regional levels) have put on a strong incentive on the implementation of the policy. Externalities can be found in term of functional externalities, which are associated with multiple and overlapping policy areas like transportation, land use planning, and housing.

We hypothesize that natural disasters have an association with climate change mitigation and adaptation policies. Solecki, Leichenko, and O'Brien (2011), for example, identify that DRR planning and management generally emerged after occurrence of a major catastrophe. Sudden natural disasters create external shocks that provide a "window of opportunity" for a local government to adopt new policies[2] (Birkland 1997). The recovery period with increased external aid offers communities opportunities for social, economic, and physical development that can reduce disaster losses in the future (Berke, Kartez, and Wenger 1993).

Extreme events can also lead to changes in political agenda at higher levels of government. For example, the WTC terrorist attacks led to the creation of the Department of Homeland Security. Similarly, Hurricane Katrina brought in legislative changes in the form of the Post-Katrina Emergency Management Reform Act of 2005. Changes in agenda at upper levels of hierarchy is expected to trickle down to lower levels of government. Legal rules and mandates are likely to reduce opportunistic behavior of actors thus motivating higher coordination levels in adoption of disaster risk reduction or climate change response strategies.

According to Birkmann et al. (2010), sudden significant changes can lead to institutional changes and propel organizations previously not involved with climate change initiatives to renegotiate on leadership factors and become involved in CCA with already engaged institutions.

Two causal mechanisms may explain the association between incidence of natural disasters and climate change mitigation and adaptation policies. First, based on the agenda setting literature, disaster events come to dominate policy agenda highlighting voluntary acceptance of responsibility by elected and appointed officials. The explanation for policy response can be drawn from the Punctuated Equilibrium Theory, which suggests that organizations evolve by going through "long periods of stability in their basic pattern of activity that are punctuated by relatively short bursts of fundamental changes" (Romanelli and Tushman 1994, 1141). Incidence of extreme events can change the perceptions of local residents regarding payoffs of coordination (Luong and Weinthal 2004) thus forming new institutions or bringing changes to existing institutional structures. In other words, policies and strategies adopted to mitigate the consequence of disasters trigger public investment in the preparation for and response to disasters.

Second, media attention explaining why the events occurred influenced public opinions. The media bring experts to explain why events might occur and change the public understanding of the problems. The media also makes policies visible by raising public awareness and consciousness. For example, Hamin and Gurran (2009) noted that the framework for adaptation tends to focus on technology and construction allowing for weather forecasting techniques, changing land use, energy, and building codes, and developing/encouraging investment in new technology to accommodate change in climate conditions. According to Birkland (1997), public attention shifts to disaster management policies and politicians who are expected to adopt disaster policies to satisfy local constituents. Since a major disaster is often regarded as a regional event, it is thus expected that local constituents of adjoining jurisdictions will also demand disaster policies from their local governments. Such demand causes a shift in policy priorities of the concerned jurisdictions. Because of coherence of new priorities developed by local actors, they are more likely to coordinate policies that have not been considered previously.

RESEARCH DESIGN

Research Site

The study was conducted in the state of Florida. The state is suitable for this study because of several reasons (Outka and Feiock 2012). First, Florida has traditionally been a "purple state" implying that it has been under the influence of varying political ideologies that have placed varying emphasis on climate change policies. Second, municipal home rule is granted in the state constitution thus allowing individual cities sufficient power to implement their own policies, thus creating much variation in policy implementation. Third, Florida also exhibits variation in city government structures like council-manager or mayor-council forms of government, thus providing variation in political/management structures of cities (Outka and Feiock 2012).

Other than the variation in political environment over the years regarding climate change, Florida also offers an important study area for smart growth because it has been a strong advocate of smart growth since the passage of the Florida Growth Management Act in 1985. Over the years, forms of smart growth implementation in Florida have changed, thus creating variation in the how smart growth is implemented. Ben-Zadok (2005) identifies that smart growth policies have been implemented in Florida in primarily three stages of development: consistency (1985–1993), concurrency (1986–1993), and compact development (1993–2002). Consistency refers to coordination between state mandates and local development plans, concurrency aims at controlled growth and economic development, and compact development focuses on creating higher density development, curbing sprawl, and economic development. It is identified that since the 3Cs as advocated by the GMA in Florida became uniform rules for all communities, making and enforcing these rules raised many tensions and conflicts.

Ben-Zadok (2005) identifies that during the consistency period of regulation; it was a more top-down approach of implementing smart growth in Florida. However, because of conflicts arising out of the situation, the state has allowed more discretion in the subsequent stages of concurrency and compact development. While discretion has introduced flexibility in the growth management process, it has also created variation in the success of these policies across the cities.

Being one of the most active states for natural disasters, Florida is also appropriate to study policy implementation and changes after a disaster because of data that can be gathered on natural disasters. In a 2013 report by FEMA, for example, Florida ranks fifth in the number of disaster declarations among U.S. states. The average amount of damages suffered by Florida per year was about $15 billion. The major types of disasters that inflict maximum losses in the state have been hurricanes, floods/flash floods, and tornadoes. Other extreme weather events prevalent in the state are thunderstorms and lightning, hail, heavy rain, and rip currents.

Sample Selection

The data to determine the association between natural disasters and smart growth policies in Florida were based on a survey that was conducted by the Florida State University (Outka and Feiock 2012). The climate change policy data from the Energy Sustainable Florida Communities research instrument captured important strategies adopted by local governments in Florida. The instrument seeks to gather information about the capacity and initiative among Florida local governments to implement sustainable climate change policies that include smart growth policies, energy efficiency, and green innovation practices. The survey also records community, political, and economic conditions that can affect implementation of these policies.

The survey was administered to 327 jurisdictions in Florida in 2009 and directed to the chief planning officers of each government. With a response rate of 50.5 percent, the sample size is 165 jurisdictions (Outka and Feiock 2012). Table 4.1 provides descriptive statistics of the sample. Based on the U.S. Census 2010, the population ranges from 463 to 391,458. Average percentage of male population is 48.15 percent. The average percentage of White population is 76.92 (comparable to the national average of 72.4 percent) with Black/African American as 16.13 percent (national average is 12.6 percent) and Hispanic as 14.36 percent (lower than the national average of 16.3 percent). The sample has a higher average for owner occupied housing units (54.46 percent) as compared to renter occupied units (24.18 percent). The household size for the former is 2.47 (in contrast to 2.65 as national average) and for the latter is 2.48 (close to the national average of 2.44). On average 71.67 percent of housing units are built after 1970, thus implying that the cities sampled have housing units in resilient condition (Cutter, Burton, and Emrich 2010). On average, few households do not own any vehicle (5.67 percent) and a small part of the population uses public transport for commuting to work (around 0.8 percent as compared to the national average of approximately 5 percent). The average time spent in commute (24.71 minutes) is also very close to the national average of 25.9 minutes. On average the labor force is 59.51 percent, which is lower than the national average of 63.7 percent. Sectors that have the highest employed population include business, manufacturing, and sales- or service-related occupation. There is a low reliance on natural resource–based industry at only 5.82 percent. The average household income is $49,826.

Table 4.1. Descriptive statistics of Florida jurisdictions in the sample

	Mean	SD	Min.	Max.
DEMOGRAPHIC CHARACTERISTICS				
Total population	35,182.42	5,851.21	463.00	391,458.00
Male (%)	48.15	3.58	28.80	59.20
Median age	42.26	8.81	24.50	70.70
White (%)	76.92	17.98	14.40	100
Black (%)	16.13	16.91	0	80.20
Hispanic (%)	14.36	14.71	0	95.60
Education				
HOUSING CHARACTERISTICS				
Total housing units	16,957.75	27,359.50	310.00	182,071.00
Housing occupancy (%)	78.64	13.87	31.60	96.90
Owner occupied (%)	54.46	14.09	19.41	91.86
Average HHS of owner occupied	2.47	0.44	1.69	4.68
Renter occupied (%)	24.18	11.31	1.79	57.97
Average HHS of renter occupied	2.48	0.56	1.24	5.30
Average HH income	49,826.37	18,371.88	21,179	123,000
Median HH income	68,567.88	29,906.84	33,331	201,784
TRANSPORTATION DATA				
No vehicle ownership (%)	5.67	3.94	21.61	0
Mean travel time (mins.)	24.71	4.32	13.9	36.7
Commuting to work using public transportation	444.90	1,748.88	0	20,250
ECONOMIC INDICATORS				
Population above 16 years in labor force (%)	59.51	10.71	19.86	77.76
Agriculture, forestry, fishing, and hunting (%)	0.81	2.20	0	18.82
Business, science, and arts (%)	18.39	6.56	3.9	39.14
Manufacturing (%)	18.39	6.56	3.9	39.14
Natural resources, construction, and maintenance (%)	5.82	3.66	0	27.21
Retail trade (%)	6.80	2.10	0.86	11.13
Sales and office occupations (%)	14.74	4.09	2.7	23.45
Service occupations (%)	10.30	3.78	0.56	25.96
Transportation-related services (%)	2.62	1.49	0	10.65
Wholesale trade (%)	1.50	1.00	0	6.04

Source: Data from U.S. Census 2010

SURVEY INSTRUMENT: ENERGY SUSTAINABLE FLORIDA COMMUNITIES (ESFC)

Climate Change Mitigation Policies

In the ESFC instrument, one question asked about local governments' climate change mitigation policies. The question queried "Has your jurisdiction established land use policies or programs to encourage or require the following?" The respondents were presented with eight types of policies or programs: (1) reduce greenhouse gas emissions, (2) compact developments in new and existing neighborhoods, (3) mixed-use development, (4) transit-oriented development, (5) infill development, (6) community-wide bicycle/pedestrian plan, (7) street design for multimodal mobility in developments, and (8) LEED or other Green development certification for new development or redevelopment projects. The respondents can answer each tool through three options: encourage, require, or no programs or policies in place.

The responses were recoded and reclassified as a dummy variable, that is, if local governments have policies that encourage and require the implementation of each of the policy categories was coded as 1, otherwise 0. Table 4.2 summarizes the response. For example, about 68.8 percent of respondents reported having no policies/program related to reduction of GHG emission, while 75.6 percent have policies that encourage or require infill development.

Table 4.2. Whether local governments in Florida have established land use policies and programs

Established Land Use Policies/Programs	Require / Encourage	No Policies / Program
Reduce greenhouse gas emission (n=128)	40 (31.25)	88 (68.75%)
Infill development (n=135)	102 (75.6%)	33 (24.4%)
LEEDs/Green cert. for new dev./ redevelopment (n=129)	48 (37.2%)	81 (62.8%)
Compact dev. in new/existing neighborhoods (n=135)	85 (63.0%)	50 (37.0%)
Transit-oriented development (n=130)	75 (57.7%)	55 (42.3%)
Comm.-wide bicycle/pedestrian plan (n=134)	84 (62.7%)	50 (37.3%)
Street design for multimodal mobility in dev. (n=132)	76 (57.6%)	56 (42.4%)
Mixed-use dev. (n=135)	111 (82.2%)	24 (17.8%)

Source: Data from Energy Sustainable Florida Communities Survey, 2009.

Climate Change Adaptation Policies

Another question in the ESFC instrument asked about climate change adaptation policy. The question asked "Does your jurisdiction's site plan and development review encourage the following green construction and technology issues?" The respondents were presented with 10 types of policies or programs: (1) daylighting, (2) certified green buildings, (3) energy-efficient buildings, (4) reduced water use, (5) heat island reduction, (6) passive and/or active solar collection, (7) on-site renewable energy sources, (8) light pollution reduction, (9) green roofs, and (10) permeable paving. The respondents can answer each tool either through a "Yes" or a "No." Table 4.3 below summarizes the response provided by local governments in our sample. The majority of local governments reported that they do not have a site plan and development review for the listed green construction and technology.

DATA ON NATURAL DISASTERS IN FLORIDA

Natural Disaster in Florida

Disaster data for this study is collected from the National Climatic Data Center (NCDC). Scholars have used different kinds of datasets to record extreme events. The presidential declarations that are managed by FEMA are

Table 4.3. Whether local governments in Florida have site plan and development review encouraging green construction and technology issues

Site Plan and Development Review Encourage the Green Construction and Technology	Yes	No
Daylighting (n=144)	7 (4.9%)	137 (95.1%)
Certified Green building (*n*=145)	29 (20.0%)	116 (80.0%)
Energy-efficient building (*n*=145)	36 (24.8%)	109 (75.2%)
Reduced water use (*n*=145)	58 (40.0%)	87 (60.0%)
Heat island reduction (*n*=145)	20 (13.8%)	125 (86.2%)
Passive and/or active solar collection (*n*=145)	10 (6.9%)	135 (93.1%)
On-site renewable energy sources (*n*=144)	7 (4.9%)	137 (95.1%)
Light pollution reduction (*n*=146)	36 (24.7%)	110 (75.3%)
Green roofs (*n*=145)	16 (11.0%)	129 (89.0%)
Permeable paving (*n*=146)	46 (31.5%)	100 (68.5%)

Source: Data from Energy Sustainable Florida Communities Survey, 2009.

one of them. Extreme events are also otherwise classified depending upon casualties or damage sustained by disaster victims (De Boer 1990), or introducing risk assessments and socioeconomic data (Cutter 2002. Using presidential declarations is, however, criticized because of political motivations behind the declaration process and hence the possibility that they might not capture the total number of extreme events occurring in a certain area. The NCDC dataset provides any extreme event and the list does not depend on whether an event has been declared as disaster or emergency. The dataset is thus expected to give a more appropriate account of the number of events in each jurisdiction and the associated damages.

In the NCDC data, extreme weather events are defined as those events that are "lying in the outermost 10 percent of a place's history." The events are classified as droughts, heat waves, snowfall, severe storms (i.e., tornadoes, hail, straight-line winds), tropical cyclones, freeze events, and winter storms. In Florida, the frequency of disaster events for each city was collected from the years 2005 to 2009. The number of extreme events during this period for the 165 jurisdictions ranged from 0 to 50 events. About 19.4 percent of the sample reported 0 events during this time period. The frequency of natural disasters was recoded as dummy, where local government experiencing a disaster between 2005 and 2009 was coded 1, otherwise 0. There are 32 (19.4 percent) local governments reporting 0 disasters and 132 (80 percent) jurisdictions reporting some form of extreme event.

Property Damage Caused by Natural Disasters

Disaster data from the National Climatic dataset is integrated to find the number of disasters that each city experienced between 2005 and 2009. We also gathered data related to the total property damage that resulted from the disasters.

An inspection of the skewness and kurtosis measures and standard errors as well as a visual inspection of the histogram showed that the data on property damages were not normally distributed. The range of property damage reported by jurisdictions varied from $0 to $68 million. A nonparametric Levene's test was performed to verify the equality of variance in the data (homogeneity of variance ($p>0.05$).

ANALYSIS AND RESULTS

Is there an association between incidence of natural disasters and climate change policies? Climate change mitigation policies appear to have an

Table 4.4. Association between incidence of natural disasters and land use policies and programs (climate change mitigation policies) adopted by local governments in Florida

Local government established land use policies/programs (expected counts)	Natural Disasters (2005–2009)		χ^2	Phi Value	p-value
	Yes	No			
REDUCE GREENHOUSE GAS EMISSION ($N=127$)					
Encourage/require	**32.1**	**7.9**			
No policies/programs	**69.9**	**17.1**	**3.464**	**0.165**	**0.063**
INFILL DEVELOPMENT ($N=134$)					
Encourage/require	**81.4**	**20.6**			
No policies/programs	**25.6**	**6.4**	**3.220**	**0.155**	**0.073**
LEED/GREEN CERT. FOR NEW DEV. OR REDEV. ($N=128$)					
Encourage/require	38.3	9.7			
No policies/programs	63.8	16.2	1.557	0.11	0.212
COMPACT DEV. IN NEW/EXISTING NEIGHBORHOODS ($N=134$)					
Encourage/require	67.2	17.8			
No policies/programs	38.8	10.2	1.484	0.105	0.223
STREET DESIGN FOR MULTIMODAL MOBILITY IN DEV. ($N=131$)					
Encourage/require	60.3	15.7			
No policies/programs	43.7	11.3	1.359	0.102	0.244
TRANSIT- ORIENTED DEVELOPMENT ($N=129$)					
Encourage/require	59.9	15.1			
No policies/programs	43.1	10.9	0.886	0.083	0.346
COMM.- WIDE BICYCLE/PEDESTRIAN PLAN ($N=133$)					
Encourage/require	66.3	17.7			
No policies/programs	38.7	10.3	0.91	0.026	0.763

Notes: Mixed-use dev. policies/programs category was excluded because expected count was less than five. **Bold** values show significant results.

association with incidence of natural disasters, especially in a policy area that is related to infill development (Table 4.4). The association is moderately low (Phi value=0.165, $p<0.10$). Local governments experiencing natural disasters tend to have policies that either encourage or require infill development [$\chi 2(1, 3.364)$, $p=0.063$].

However, a higher number of local governments experiencing natural disasters tend not to have a mitigation policy related to a reduction in GHG emission. In other words, incidence of natural disasters do not seem to be associated with local governments adopting climate change mitigation policies in the form of reduction of GHG emission (Table 4.4).

Table 4.5. Property damage differences and land use policies and programs (climate change mitigation policies) adopted by local governments in Florida

| Whether local government established land use policies/programs | Property Damages* (2005–2009) | | | |
	Mean	Mean Diff.	t-value**	p-value
REDUCE GREENHOUSE GAS EMISSION				
Encourage/require (n=40)	$2,120,693			
No policies/programs (n=87)	$400,537	$1,720,156	3.047	0.003
INFILL DEVELOPMENT				
Encourage/require (n=102)	$1,153,220			
No policies/programs (n=32)	$73,031	$1,080,189	1.136	0.258
LEED/GREEN CERT. FOR NEW DEV./REDEVELOPMENT				
Encourage/require (n=48)	$2,203,677			
No policies/programs (n=80)	$172,134	$2,031,542	3.579	0.000
COMPACT DEV. IN NEW/EXISTING NEIGHBORHOODS				
Encourage/require (n=85)	$1,239,247			
No policies/programs (n=49)	$298,500	$940,747	2.605	0.010
STREET DESIGN FOR MULTIMODAL MOBILITY IN DEVELOPMENT				
Encourage/require (n=76)	$1,426,611			
No policies/programs (n=55)	$204,036	$1,222,575	1.270	0.206
TRANSIT-ORIENTED DEVELOPMENT				
Encourage/require (n=75)	$1,522,140			
No policies/programs (n=54)	$105,032	$1,417,107	1.200	0.232
COMM.-WIDE BICYCLE/PEDESTRIAN PLAN				
Encourage/require (n=84)	$1,243,324			
No policies/programs (n=49)	$414,790	$838,533	0.735	0.464

Notes: *Property damage data virtually inspected with histograms indicating they were not normally distributed.
**Nonparametric tests were performed after testing for Levene's equality of variance.
Bold values represent significant results.

There is also significant difference in property damage from natural disasters between local governments that encourage/require LEED/Green certification for new development or redevelopment and those who do not have any policies for the same (Table 4.5). Property damage was found to be significantly higher for those who encouraged/required such policies. The significance of the difference is high (t=3.579, p=0.000). There is also evidence to suggest difference in property damage from natural disasters is statistically significant between jurisdictions who encourage/require policies

Table 4.6. Association between incidence of natural disasters and climate change adaptation policies and programs adopted by local governments in Florida

Site Plan and Development Review Encourage Green Construction and Technology (Expected counts)	Natural Disasters (2005–2009)		χ^2	Phi Value	p-value
	Yes	No			
Certified Green building (n=145)					
Yes	23.2	5.8			
No	92.8	23.2	0.172	0.034	0.678
Energy-efficient building (n=145)					
Yes	28.8	7.2			
No	87.2	21.8	0.748	0.072	0.387
Reduced water use (n=145)					
Yes	46.4	11.6			
No	69.6	17.4	2.328	0.127	0.127
Light pollution reduction (n=146)					
Yes	28.8	7.2			
No	88.2	21.8	0.167	0.034	0.683
Permeable paving (n=146)					
Yes	36.9	9.1			
No	80.1	19.9	0.004	0.005	0.951

Notes: Daylighting, heat island reduction, passive/active solar collection, and on-site renewable energy sources categories were excluded from final analysis.

on compact development in new/existing neighborhoods and those who do not have any such policies in place, that is, t=2.065, p=.01 (see Table 4.5).

What are the associations between climate change adaptation policies and natural disasters? Table 4.6 shows that there is no evidence to suggest a statistically significant association between occurrence of natural disasters and climate change adaptation policies, that is, (1) certified green building, (2) energy-efficient building, (3) reduced water use, (4) light pollution reduction, and (5) permeable paving. However, when property damage resulting from natural disasters is considered, we found a significant difference in property damage from natural disasters between local governments who adopted and those that did not adopt policies for reduced water usage. Property damage was found to be significantly higher for those adopting reduced water usage policy (t=1.993, p<.05) (Table 4.6).

A significant difference between property damage from natural disasters was also seen between local governments who adopt and do not adopt

Table 4.7. Association between property damage differences and green construction and technology (climate change adaptation policies) adopted by local governments in Florida

Site Plan and Development Review Encourage Green Construction and Technology	Property Damages* (2005–2009)			
	Mean	Mean Diff.	t-value**	p-value
Certified Green building				
Yes (n=29)	$2,787,758			
No (n=116)	$398,299	$2,389,459	0.898	0.371
Energy-efficient building				
Yes (n=36)	$2,257,041			
No (n=109)	$420,130	$1,836,910	0.064	0.949
Reduced water use				
Yes (n=58)	$1,874,086			
No (n=87)	$210,928	$1,663,158	1.993	**0.048**
Light pollution reduction (N=146)				
Yes (n=36)	$2,649,951			
No (n=110)	$288,606	$2,361,344	0.374	0.709
Permeable paving (N=146)				
Yes (n=46)	$872,565			
No (n=100)	$869,107	$3,457	1.674	**0.096**

Notes: *Property damage data virtually inspected with histograms indicating they were not normally distributed.
**Nonparametric tests were performed after testing for Levene's equality of variance (Nordstokke et al. 2011).
Bold values represent significant results.

policies regarding permeable paving (Table 4.7). Property damage was significantly higher for local governments who adopted tools for permeable paving (t=1.674, p<.1).

DISCUSSION

Based on data collected among 165 municipalities in Florida, the results of our analysis suggest that climate change mitigation policies appear to have an association with incidence of natural disasters. In particular, the association is significant in a policy area that is related to infill development. We also found that local governments that did not require/encourage

GHG emission policies tend to have a lower amount of property damages compared to those that have policies related to reduction in GHG emission. Similar results on property damages can be found among local governments with compact development policy and LEED/Green certification for new development/redevelopment policy. As for the association between climate change adaptation policy and natural disasters, we found that a significant difference in property damages between local governments with mechanism to encourage green construction and technology related to reduced water use and permeable paving.

How to make sense of the analysis and results based on the Florida dataset? In other words, why are property damages (caused by natural disasters) linked to prevalence of mitigation policies? One possible explanation is that a higher amount of damage would encourage local governments to adopt mitigation policies such as reduction of GHGs emissions, Green certification for new development/redevelopment, and compact development in new/existing neighborhoods. These policies aim at reducing expansion of urban area and reduction of vehicular transit.

The association between natural disasters and climate change mitigation policies (as manifested through the smart growth policies) can also be attributed to the fact that with increases in tangible losses from natural disasters, local governments become more cognizant of the economic risks associated with climate change. Hence, in order to mitigate future risks from disasters, policies that help in reducing GHG emissions—one of the most prominent factors leading to climate change—are adopted. Compact development, for example, helps in restricting urban growth to open areas that might be more vulnerable to disasters, thus such policy can mitigate future risks.

However, other smart growth policies such as transit-oriented development or pedestrian-friendly street design, though also indirectly associated with reducing GHG emissions, are generally linked to enhancing quality of life of local communities and hence are not directly associated with risk mitigation. This might explain why we might not find a significant difference in property damage between local governments that encourage/require the provision for these latter policies than those that do not.

We also found that the incidence of natural disasters is associated with infill development policies. Infill development, for example, helps in restricting spatial growth of urban areas by utilizing vacant lands within existing communities for new development. However, Farris (2001) suggests that urban infill (one of the tools of Smart Growth) face potential market problems. Emphasis on infill development and restricting new development limit lifestyle choices of contemporary urban families, affect land prices and property values in suburban regions and homeownership rates.

Based on our analysis, it can be suggested here that, with occurrence of disasters, local governments adopt policies that restrict development in new hazardous areas that will increase future risk to disasters. Infill/compact development policies also aim at efficient utilization of infrastructure and natural resources. Redevelopment and increasing densities within existing neighborhoods promotes usage of existing infrastructure networks like water and sewerage systems instead of building anew when vacant/new land is acquired for development. Hence optimum utilization of existing resources also helps in mitigating climate change factors, thus motivating local governments to implement these policies.

The results indicate that there is no significant association between incidence of disasters and adoption of climate change adaptation policies. However, property damage resulting from natural disasters does have a significant association for adopting reduced water usage and permeable paving. Both of the adaptation policies aim at managing water resources. Permeable paving helps in reducing runoff and maintaining groundwater levels. The reason for the significant association with the mentioned tools and no evidence for other adaptation tools is not clear. Water saving policies in Florida may have been already in place due to state water shortages or other area-specific factors. Hence further investigation is required to understand the association between occurrence of disasters and adoption of climate change adaptation policies.

CONCLUSION

This chapter examines whether there is an association between local governments' decision to adopt smart growth and the incidence of natural disasters. The results indicate that climate change mitigation tools like compact development and infill development are significantly associated with property damage resulting from disasters. However, there is very less evidence of significant association between disaster incidence and climate change adaptation tools. The association of disaster incidence and adoption of climate change adaptation tools is not very clear and needs to be looked into further.

The study is based upon a sample size of 165 cities in Florida. Though the average demographic characteristics of the sample is representative of U.S.

cities, there are limitations. Florida is a coastal state and is one of the most active regions in the U.S. for a number of extreme events like hurricanes, floods and thunderstorms. Hence the state can be more proactive in adoption of climate change response policies. The state also has a long-standing tradition of smart growth policies. Such tools may not be prevalent for other states that do not prefer compact development. The study may also be limited in the context of other forms of disasters like drought, winter storms, and so forth that do not occur or are not prevalent in Florida.

However, despite the limitations of the study, the chapter has both theoretical and practical contributions. Theoretically, it contributes toward the utilization a framework in smart growth policy area and also in the context of disasters. We combine elements from the punctuated equilibrium theory and agenda-setting literature. As for practice, smart growth covers a wide range of tools, and each tool has a different implication for climate change response policies. Each tool also has unique challenges for implementation. This chapter hence contributes toward categorizing smart growth in terms of climate change response policies and also will give practitioners a preliminary idea of what tools cities are more likely to adopt in the context of disasters.

Notes

1. Political opposition to smart growth approach in building a resilient city is often associated with restriction of urban sprawl. The argument has been regarded as anti-growth, especially legislation and land-use policies restricting property rights and the enforcement of environmental preservation related to open spaces and land-use control. Smart growth policy adoption hence often faces oppositions and need adjustments of conflicting priorities.

2. Extreme events can also lead to changes of political agendas at higher levels of government. The WTC terrorist attacks led to the formation of the Department of Homeland Security. Hurricane Katrina brought in legislative changes in the form of the Post-Katrina Emergency Management Reform Act, 2005. Changes in such hierarchical agendas resulting from extreme events is expected to cause changes in coordination. Legal mandates or rules can restrict opportunistic behavior of actors, thus reducing higher transaction cost risks associated with coordination and thus motivating higher coordination levels among actors.

References Cited

American Planning Association. 2012. "APA Policy Guide on Smart Growth." Last accessed February 14, 2020. https://www.planning.org/policy/guides/adopted/smartgrowth.htm.

Barbour, E., and M. B. Teitz. 2001. "A Framework for Collaborative Regional Decision-Making." San Francisco: Public Policy Institute of California.

Berke, P. R., J. Kartez, and D. Wenger. 1993. "Recovery After Disaster: Achieving Sustainable Development, Mitigation and Equity." *Disasters* 17(2): 93–109.

Ben-Zadok, E. 2005. "Consistency, Concurrency and Compact Development: Three Faces of Growth Management Implementation in Florida." *Urban Studies* 42(12): 2167–2190.

Birkland, Thomas A. 1997. *After Disaster: Agenda Setting, Public Policy, and Focusing Events*. Washington, D.C.: Georgetown University Press.

Birkmann, J., P. Buckle, J. Jaeger, M. Pelling, N. Setiadi, M. Garschagen, N. Fernando, and J. Kropp. 2010. "Extreme Events and Disasters: A Window of Opportunity for Change? Analysis of Organizational, Institutional and Political Changes, Formal and Informal Responses after Mega-Disasters." *Natural Hazards* 55(3): 637–655.

Brown, M. A., and F. Southworth. 2008. Mitigating Climate Change Through Green Buildings and Smart Growth. *Environment and Planning A* 40(3): 653–675.

Burby, R. J. 2005. "Have State Comprehensive Planning Mandates Reduced Insured Losses from Natural Disasters?" *Natural Hazards Review* 6(2): 67–81.

Burby, R. J., R. E. Deyle, D. R. Godschalk, and R. B. Olshansky. 2000. "Creating Hazard Resilient Communities Through Land-Use Planning." *Natural Hazards Review* 1(2): 99–106.

Burby, R. J., A. C. Nelson, D. Parker, and J. Handmer. 2001. "Urban Containment Policy and Exposure to Natural Hazards: Is There a Connection?" *Journal of Environmental Planning and Management* 44(4): 475–490.

Coaffee, J. 2008. "Risk, Resilience, and Environmentally Sustainable Cities." *Energy Policy* 36(12): 4633–4638.

Cutter, S. L., C. G. Burton, and C. T. Emrich. 2010. "Disaster Resilience Indicators for Benchmarking Baseline Conditions." *Journal of Homeland Security and Emergency Management* 7(1).

Cutter, Susan L., ed. 2002. "American Hazardscapes: The Regionalization of Hazards and Disasters." Washington, D.C.: Joseph Henry Press.

Dawkins, C. J., and A. C. Nelson. 2002. "Urban Containment Policies and Housing Prices: An International Comparison with Implications for Future Research." *Land Use Policy* 19(1): 1–12.

De Boer, J. 1990. "Definition and Classification of Disasters: Introduction of a Disaster Severity Scale." *Journal of Emergency Medicine* 8(5): 591–595.

de Oliveira, J.A.P. 2009. "The Implementation of Climate Change Related Policies at the Subnational Level: An Analysis of Three Countries." *Habitat International* 33(3): 253–259.

Farris, J. T. 2001. "The Barriers to Using Urban Infill Development to Achieve Smart Growth." *Housing Policy Debate* 12(1): 1–30.

Feiock, R. C. 2013. "The Institutional Collective Action Framework." *Policy Studies Journal* 41(3): 397–425.

Luong, P. J., and E. Weinthal. 2004. "Contra Coercion: Russian Tax Reform, Exogenous Shocks, and Negotiated Institutional Change." *American Political Science Review* 98(1): 139–152.

Frank, L. D. 2000. "Land Use and Transportation Interaction: Implications on Public Health and Quality Of Life." *Journal of Planning Education and Research* 20(1): 6–22.

Hamin, E. M., and N. Gurran. 2009. "Urban Form and Climate Change: Balancing Adaptation and Mitigation in the US and Australia." *Habitat international* 33(3): 238–245.

Larsen, K., and U. Gunnarsson-Östling. 2009. "Climate Change Scenarios and Citizen-Participation: Mitigation and Adaptation Perspectives in Constructing Sustainable Futures." *Habitat International* 33(3): 260–266.

Laukkonen, J., P. K. Blanco, J. Lenhart, M. Keiner, B. Cavric, and C. Kinuthia-Njenga. 2009. "Combining Climate Change Adaptation and Mitigation Measures at the Local Level." *Habitat International* 33(3): 287–292.

Nordstokke, D. W., B. D. Zumbo, S. I. Cairns, and D. H. Saklofske. 2011. "The Opening Characteristics of the Nonparametric Levene Test for Equal Variances with Assessment and Evaluation Data." *Practical Assessment, Research & Evaluation* 16(5): 1–8.

O'Brien, G., P. O'keefe, J. Rose, and B. Wisner. 2006. "Climate Change and Disaster Management." *Disasters* 30(1): 64–80.

O'Connell, L. 2009. "The Impact of Local Supporters on Smart Growth Policy Adoption." *Journal of the American Planning Association* 75(3): 281–291.

Outka, U., and R. C. Feiock. 2012. "Local Promise for Climate Mitigation: An Empirical Assessment." *William & Mary Environmental Law and Policy Review* 36: 635–670.

Park, S. C., J. W. Park, and K. H. Lee. 2012. "Growth Management Priority and Land-Use Regulation in Local Government: Employing a Full Structural Equation Model." *International Review of Public Administration* 17(1): 125–147.

Ramirez de la Cruz, E. E. 2009. "Local Political Institutions and Smart Growth: An Empirical Study of the Politics of Compact Development." *Urban Affairs Review* 45(2): 218–246.

Romanelli, E., and M. L. Tushman. 1994. "Organizational Transformation as Punctuated Equilibrium: An Empirical Test." *Academy of Management Journal* 37(5): 1141–1166.

Schipper, L., and M. Pelling. 2006. "Disaster Risk, Climate Change and International Development: Scope for, and Challenges to, Integration." *Disasters* 30(1): 19–38.

Shapiro, J. M. 2006. "Smart Cities: Quality of Life, Productivity, and the Growth Effects of Human Capital." *Review of Economics and Statistics* 88(2): 324–335.

Solecki, William, Robin Leichenko, and Karen O'Brien. 2011. "Climate Change Adaptation Strategies and Disaster Risk Reduction in Cities: Connections, Contentions, and Synergies." *Current Opinion in Environmental Sustainability* 3(3): 135–141.

5

Can Transportation Be Sustainable in the Suburbs?

A Case Study of Complete Streets on Long Island, New York

SANDRA J. GARREN, MATHEW K. HUXEL, AND CAROLINA A. URREA

Smart Growth and Complete Streets movements have the potential to transform American suburbs with thriving sustainable transportation systems, but what is sustainable transportation? There is no universally agreed-upon definition for sustainable transportation (Steg and Gifford 2005; Zhou 2012). However, sustainable transportation is often described in terms similar to the definition of sustainable development advanced by the Brundtland Commission (Steg and Gifford 2005). Namely, a sustainable transportation system provides mobility that does not limit economic development or harm the environment while at the same time meeting the needs of all members of society. What we have today in American can largely be characterized as unsustainable transportation with a dominant car culture that results in environmental problems such as air pollution and climate change. The current system poses a risk to people due to injuries from traffic accidents, increased incidences of asthma from air pollution, and excessive noise pollution. In addition, the system does not meet the needs of all of its citizens (i.e., the disabled, children, and seniors), and for some, car ownership is out of reach due to economic factors.

However, there are examples of more sustainable transportation systems in urban cities such as New York, Chicago, and San Francisco which have compact development and economies of scale that support public transit. Globally, Sustainable Development Goal (SDG) 11 (see chapter 1) is focused on making cities more inclusive, safe, resilient, and sustainable. SDG Target 11.2 specifically focuses on providing access to public transportation particularly for vulnerable populations. In contrast to urban cities, sustainable transportation in the suburbs has lagged for a variety of reasons most notably from development patterns that have favored automobiles over public transit. There is evidence that this pattern is changing with over 90 U.S. cities and communities being designed with sustainable transportation under Leadership for Energy and Environmental Design (LEED) for Cities and LEED for Communities programs (United States Green Building Council undated). In these certification systems, points are awarded for many design features that include access to quality transit, provide bicycle facilities, reduce parking footprint, and reduce commuting distances and greenhouse gas emissions.

Other solutions include switching out fossil fuel cars with electric cars; however, this solution addresses air pollution and climate change, but not the access issue. Ridesharing (e.g., Lyft and Uber) and carpooling programs are also a part of the solution but do not solely make networks more sustainable. Likely the most promising sustainable transportation solution is to design or redesign the urban and suburban landscapes to create communities that include mixed-use transit-oriented development (TOD) and multimodal transportation (i.e., mass transit, automobiles, bicycles, and pedestrians) according to Smart Growth principles and according to Complete Streets principles(an offshoot of Smart Growth).

This chapter examines sustainable Smart Growth and Complete Streets policies in New York and across Long Island, reviews Long Island's transportation system, and focuses on an early-adopter community, the Village of Great Neck Plaza, that has demonstrated successful sustainable transportation projects. The chapter ends by presenting lessons learned and continued challenges for the advancement of sustainable transportation on Long Island and in the suburbs in general.

Smart Growth and Complete Streets Background

Smart Growth is a land development theory that started in the late 1970s and is founded on ten basic principles of Smart Growth presented in Table 5.1 (Smart Growth America undated). At its core are mixed-use developments that are walkable (Principle 4), foster a strong sense of place (Principle 5), and include a variety of transportation choices (Principle 8). Complete Streets refers to a collection of urban design principles meant to increase the safety and accessibility of city and town streets for all users, regardless of age or physical ability and regardless of mode of transportation. Complete Streets principles attempt to increase the usability for pedestrians, cyclists,

Table 5.1. Smart Growth and Complete Streets principles

Smart Growth Principles (Smart Growth America)	Complete Streets Policy Principles (National Complete Streets Coalition)	Complete Streets Design Principles (Sharpin, Welle, and Luke 2017)
Mix land uses.	Includes a vision for how and why the community wants to complete its streets.	An active streetscape
Take advantage of compact design.		Pedestrian-scale lighting
Create a range of housing opportunities and choices.	Specifies that "all users" includes pedestrians, bicyclists, and transit passengers of all ages and abilities, as well as trucks, buses, and automobiles.	Green infrastructure Street furniture Bicycle facilities Signage
Create walkable neighborhoods.		Accessibility features for all surface types
Foster distinctive, attractive communities with a strong sense of place.	Applies to both new and retrofit projects, including design, planning, maintenance, and operations, for the entire right of way.	
Preserve open space, farmland, natural beauty, and critical environmental areas.	Makes any exceptions specific and sets a clear procedure that requires high-level approval of exceptions.	
Direct development toward existing communities.		
Provide in advance a variety of transportation choices.	Encourages street connectivity and aims to create a comprehensive, integrated, connected network for all modes.	
Make development decisions predictable, fair, and cost effective.	Is adoptable by all agencies to cover all roads.	
Encourage community and stakeholder collaboration in development decisions.	Directs the use of the latest and best design criteria and guidelines while recognizing the need for flexibility in balancing user needs.	
	Directs that Complete Streets solutions will complement the context of the community.	
	Establishes performance standards with measurable outcomes.	
	Includes specific next steps for implementation of the policy.	

motorcycles, cars, buses, or other users (National Complete Streets Coalition undated). The goal of implementing Complete Streets design principles is to reduce traffic congestion and collisions, which has the benefits of saving costs on damages to personal or public property, as well as reducing personal injury and increasing accessibility to all members of the community. When done well, Complete Streets upgrades can be part of downtown revitalization projects, produce jobs, increase foot traffic and business patronage, and some studies even suggest the economic growth and cost savings on damage and personal injury can exceed the cost of a Complete Streets project in only a few years (Anderson et al. 2015).

The largest organization promoting the policy and design principles of Complete Streets is the National Complete Streets Coalition (National Complete Streets Coalition undated) which is a part of Smart Growth America, a nonprofit organization dedicated to the process of strategically building smarter, safer, and more responsible towns and cities (Smart Growth America undated). The National Complete Streets Coalition (Coalition) was launched in and is the authority on Complete Streets policy in America (National Complete Streets Coalition 2019). The Coalition details 10 essential elements to ideal Complete Streets policy as presented in Table 5.1 (National Complete Streets Coalition 2018).

In order for a policy to be accepted as a Complete Streets Policy by the National Complete Streets Coalition, it must adhere to the 10 principles laid out in Table 5.1; however, these policy principles do not constitute detailed design principles a city planner and engineering firm can use when planning a street upgrade. The design principles vary from municipality to municipality, but there are some basic goals that are common in many Complete Streets designs. An overarching set of eight design principles (Table 5.1) were advanced by Sharpin, Welle, and Luke (2017). These eight elements encompass beautification, environmental sustainability, and accessibility, and enable any user regardless of ability or mode of transportation equal access to a city street.

Complete Streets policies and design practices represent a relatively new set of principles by which to develop transportation systems that work for all users, not just the driver, while at the same time creating community, encouraging an active lifestyle, and greening the environment.

Long Island's Unique History and Current Transportation System

Long Island has a unique history with the first mass-produced suburb in America. This suburb, Levittown, provided a model that in many ways set the trajectory of the current suburban form and reliance on the automobile not only on Long Island but across the nation. A logical place to begin this case study is to examine the historical development of Long Island's

Figure 5.1. Railroad station, Port Washington (1903). Image courtesy of the Port Washington Public Library.

transportation system and the unique characteristics that developed as a result of this development.

Long Island, with the exception of Brooklyn, consisted mainly of small villages connected by carriage routes until the Long Island Railroad (LIRR) began construction in the 1830s. The LIRR was originally designed to connect New York to Boston and was not originally built to serve the local population (Long Island Railroad History undated). The connection was never completed, and when the Connecticut branch connected Boston to New York, the LIRR shifted its focus to serve New York City commuters. The LIRR tracks were laid in segments, and by 1900, the current train lines and stations were complete. The first form of multimodal transportation on Long Island, as shown in Figure 5.1, were horses and buggies, used to pick up travelers arriving by rail.

There are currently 125 train stations, and for nearly 50 years, the LIRR was very popular and communities began to spring up near train stations (Figure 5.2a). After World War II, train usage significantly declined due to the rise of the automobile and the expansion of the road network (Figure 5.2b). As already discussed, the first mass-produced suburb, Levittown, was built in 1947, and the suburban form began to spread across the island (Long Island Railroad History undated). Suburbanization, supported by the

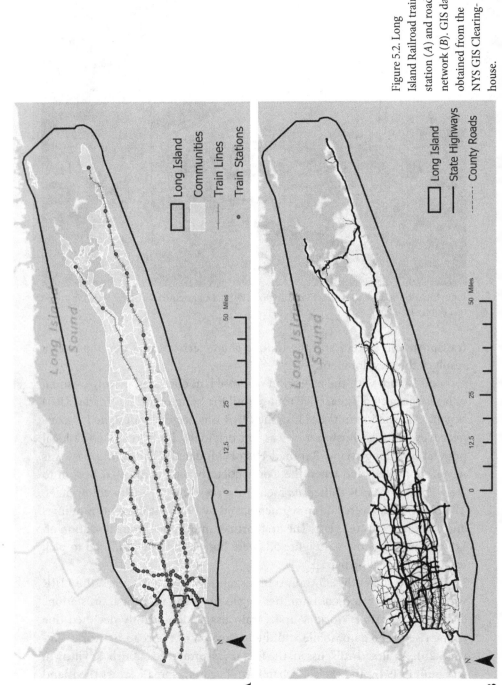

Figure 5.2. Long Island Railroad train station (A) and road network (B). GIS data obtained from the NYS GIS Clearinghouse.

construction of the Long Island Expressway (LIE) which began development in the 1940s, cemented the car as the dominant form of transportation. In 1965, New York State bought the LIRR as they were struggling under private ownership, and the LIRR is now a part of the Metro Transportation Authority (MTA) which is linked to the New York City subway system and other regional rail lines (e.g., Amtrak and the Metro-North Railroad which connects to upstate New York, New Jersey, and Connecticut). Today, the train remains an important mode of transportation for east-west commuters to New York City which is easily seen in Figure 5.2a.

Due to its history of main street development occurring near train stations in the 1800 and 1900s and suburban development beginning in the mid-1940s, Long Island has developed a collection of unique cities, villages, and hamlets that cover the entire island (Figure 5.3a). There are in total 292 uniquely named communities (i.e., two cities, 96 villages, and 194 unincorporated census designated places [CDPs]). Two more layers of government exist on Long Island that include towns (13 in total) and counties (i.e., Nassau and Suffolk Counties) (see Figure 5.3a). Urban developments include a variety of forms (e.g., main street, shopping malls, suburban residential developments, mixed use, and agricultural/ecotourism destinations). In general, the urbanization pattern on Long Island reveals more urban development in Nassau County to the west and closer to New York City and more residential suburbs followed by some rural and agricultural development eastward (Figure 5.3b).

Walkability and bikeability are two forms of transportation promoted in both Smart Growth and Complete Streets. Long Island has a reputation of having the region's most dangerous roads (Tri-State Transportation Campaign 2016). As shown on Figure 5.4a, the most walkable areas are on the western side of Long Island closest to New York City; however, walkable places are also shown in regions that have street development and that surround train stations (Figure 5.4b). It is clear from Figure 5.4b that the bike network is not well developed in Nassau and on Long Island; bike trails exist mostly for recreational use and not commuting. Aside from cars and trains, bus transit is provided by the Nassau-Inter-County Express (NICE) and Suffolk County Transit (Figure 5.5a) with annual ridership of approximately 30 million and 6 million a year, respectively (New York Metropolitan Transportation Council 2017). The bus system has experienced a decline in service, ridership funding, and ridership (TSTC 2017). Long Island has one international airport (MacArthur Airport) and two very busy international airports just west of Long Island's boundary (John F. Kennedy and

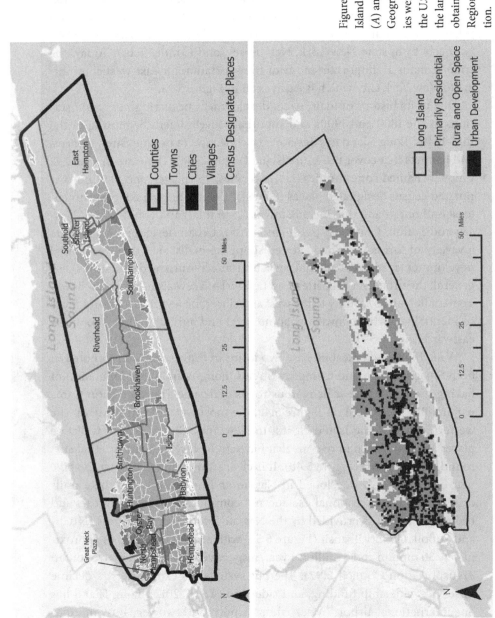

Figure 5.3. Long Island communities (A) and land uses (B). Geographic boundaries were obtained from the U.S. Census and the land use GIS data obtained from the Regional Plan Association.

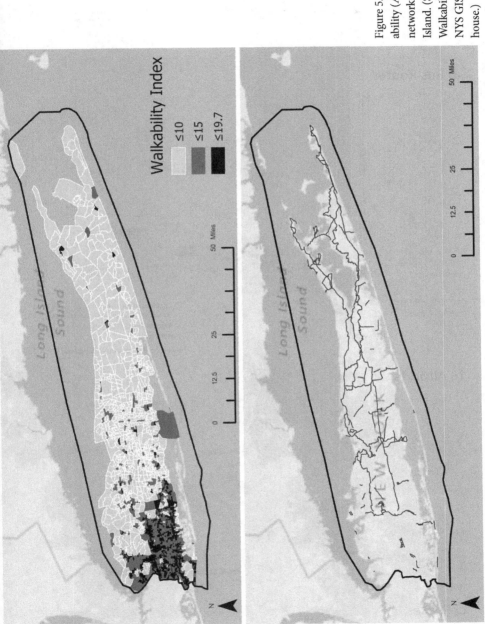

Figure 5.4. Walkability (*A*) and bicycle network (*B*) on Long Island. (Source: EPA Walkability Index and NYS GIS Clearinghouse.)

a) Bus Routes

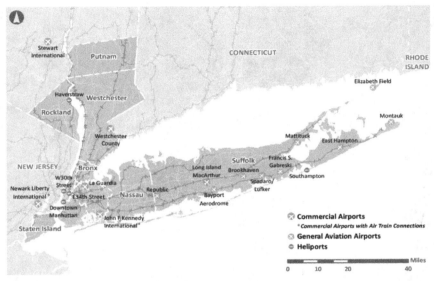

b) Airports

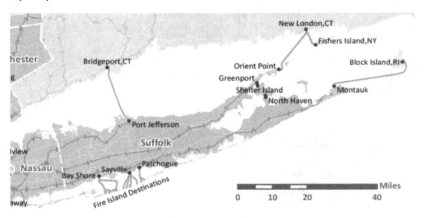

c) Ferries

Figure 5.5. Bus routes (*a*), airports (*b*), and ferries (*c*) on Long Island. (Source: New York Metropolitan Transportation Council 2007.)

La Guardia International Airports) (Figure 5.5b). Ferry service connects to Connecticut and to destination islands (Fire Island on the south shore and Shelter and Block Islands on the east end of Long Island [Figure 5.5c]).

Approximately 2.86 million people live on Long Island (Table 5.1). The Long Island population has a relatively higher median household income, lower poverty rate, and is more educated than the rest of the state and nation. Fewer people use public transit and more people commute with one driver on Long Island than in the state but Long Islanders are about average when compared to the rest of the U.S. There are more single-family residences, and the median home value is higher on Long Island than in the state and the nation; however, people on Long Island pay more for housing than elsewhere. Long Island residents enjoy a lower crime rate than the rest of the U.S.

Approximately 100,000 automobile accidents occur each year on Long Island injuring on average about 41,000 people and killing approximately 250 people over the last 17 years (NYSDOT undated). Crash data trends shown in Figure 5.6 indicate that a fairly steady rate for total and bicycle accidents and accidents that involved pedestrians, but a decreasing trend in injuries, property damage, and bike accidents.

COMPLETE STREETS POLICY

The National Complete Streets Coalition (2019), an initiative of Smart Growth America, maintains a database of Complete Streets policies across the United States. The first state to adopt a Complete Streets Policy was the state of Oregon in 1971. This policy has grown over the years, and as of August 2019, there were over 1,400 policies representing 33 state governments (Figure 5.7).

In 2010, the state of New York passed the Smart Growth Public Infrastructure Policy Act (NYSDOT 2010) with the intent to reduce the cost of sprawl and to require that public infrastructure projects adhere to Smart Growth principles (see Table 5.1). Not long after, the Complete Streets Act was signed into state law in 2011. This act requires that state, county, and local agencies consider the mobility of all users in order to receive state and federal funding for transportation projects (New York State Department of Transportation undated).

According to the National Complete Streets Coalition's (2019) inventory of Complete Streets Policy, 12 states, including New York, and 62 counties have policies, including Nassau and Suffolk Counties on Long Island. The

Figure 5.6. Long Island automobile crash trends (2000–2017). (Source: NYS Department of Transportation Statistics.)

two counties passed resolutions in 2013 and 2012, respectively. In New York, 112 towns, cities, and villages have either a resolution, policy, plan, design manuals, and/or legislation. New York City passed the first set of policies (i.e., Sustainable Streets Strategic Plan in 2008 followed by a design manual in 2009).

On Long Island, most of the south shore and a few villages and cities have enacted Complete Streets policies as shown on Figure 5.8. Six of the eleven towns are specifically addressing Complete Streets (Babylon policy in 2010; Brookhaven resolution in 2010; East Hampton resolution in 2011; Islip resolution in 2010; Southampton policy; and, North Hempstead policy in 2011). Three villages and the City of Long Beach have policies (Great Neck Plaza policy in 2012, Hempstead policy in 2012, and Valley Stream resolution in 2013) (National Complete Streets Coalition 2019). Great Neck Plaza was an

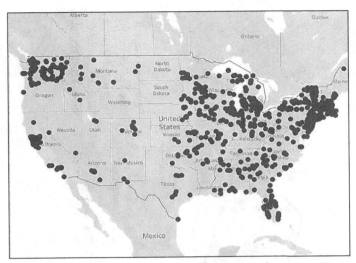

Figure 5.7. Complete Streets policy in the United States. (Source: National Complete Streets Coalition 2018.)

early adopter and has developed policies and to this date completed seven sustainable transportation projects. It is for these reasons that Great Neck Plaza was selected for a case study.

Complete Streets Case: The Village of Great Neck Plaza

First settled by Native Americans, the Great Neck peninsula has gone through transitions with the Village of Great Neck Plaza now housing families of all backgrounds and nationalities (Long Island Traditions and Village of Great Neck Plaza 2003). The 1800s expansion of the Flushing Northside Railroad, Queens Boulevard (Jericho Turnpike), and Northern Boulevard enticed investors to buy land and created an influx of visitors from Manhattan (Historic Preservation Commission of Great Neck Plaza undated). Construction of the Robertson Building in the late 1800s created storefront space, apartments, a general store, and movie theater and would become a popular gathering spot for locals and the theater crowd during the 1900s (Historic Preservation Commission of Great Neck Plaza undated). By the 1920s, Great Neck Plaza became a destination for the wealthy with the addition of the new train station and construction of apartments for upper-class city dwellers that included doctors and businessmen (Long Island Traditions and Village of Great Neck Plaza 2003). Great Neck Plaza is part

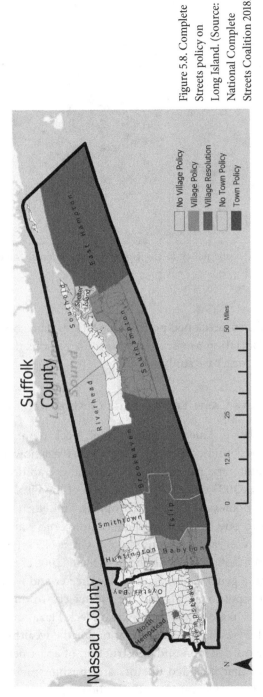

Figure 5.8. Complete Streets policy on Long Island. (Source: National Complete Streets Coalition 2018.)

of what was known as the Gold Coast which was the setting in *The Great Gatsby* with 22 historical stopping points featured in the Great Neck Plaza Historic Walking Tour (Long Island Traditions and Village of Great Neck Plaza 2003).

The Village of Great Neck Plaza was incorporated in April 1930 (Village of Great Neck Plaza undated) with the LIRR track lowered in 1935 to improve automobile and pedestrian safety at the train crossing (Long Island Traditions and Village of Great Neck Plaza 2003). While projects were delayed during the stock market crash and the Depression, investors did not abandon the village, and with federal government support the village became an example of affordable housing (Historic Preservation Commission of Great Neck Plaza undated) and both the median income and home values are lower than the rest of Long Island (Table 5.2).

By the 1950s, the village contained numerous apartments and buildings of distinct facades, sizes and amenities reflecting the fashion of the time (Long Island Traditions and Village of Great Neck Plaza 2003). The Village of Great Neck Plaza began preserving its vast history and architecture in 1976 and continued to with the creation of the Historic Preservation Commission in 1998 (Historic Preservation Commission of Great Neck Plaza undated). In 2004, Great Neck Plaza received the great distinction of being named a "Preserve America" community by the White House and Advisory Council on Historic Preservation (Advisory Council on Historic Preservation undated). The Incorporated Village of Great Neck Plaza has been able to preserve its original historic buildings while embracing the future and developing into a modern suburban community. It encompasses a busy commercial district, parks, residential sector, and a vibrant downtown with over 260 retail stores, service establishments, multifamily apartment buildings, and hotels (Village of Great Neck Plaza undated).

Great Neck Plaza is different than the rest of Long Island in a few other key areas (see Table 5.2). Approximately 82 percent in Great Neck Plaza reside in apartment buildings with 20 or more units (compared to less than 10 percent in the rest of Long Island). The population density of 22,000 people per square mile mimics densities in large urban areas like New York City (27,751 people). This density is much higher than the approximate 4,800 and 1,637 people in Nassau and Suffolk Counties, respectively. There are about twice the number of seniors (32 percent of population), more disabled (14 percent), and about half as many children (14 percent) than the rest of Long Island. And approximately 37 percent of residents use public transportation

Table 5.2. Complete Streets population, commuting, housing, and crime characteristics data in the United States, New York, and Long Island (2017)

Population characteristics	United States	State	Nassau County	Suffolk County	Great Neck Plaza
Total People	321,004,407	19,798,228	1,363,069	1,497,595	6,967
Density per square mile	87	416	4,766	1,637	22,000
Median household income	$57,652	$62,765	$105,744	$92,838	$74,303
Poverty rate	14.6%	15.1	5.9%	7.2%	6.7%
Percent non-white	38.5%	44.1%	38.6%	31.5%	38%
Percent children	22.9%	21.2%	21.9%	22%	14%
Percent seniors	14.9%	15.2%	16.8%	15.6%	31.9%
Percent with disability	12.6%	11.4%	8.5%	9.5%	14.3%
Educated (bachelor or more)	30.9%	35.3%	44.4%	35%	59.1%
COMMUTING CHARACTERISTICS					
Public transit	5.1%	28.2%	16.7%	6.4%	36.8%
Drove alone	76.4%	52.9%	68.9%	79.7%	41.7%
No car in household	4.4%	21.9%	3.3%	2.2%	15.7%
Mean commute (minutes)	26.4	33.0	35.8%	32.0%	36.2%
HOUSING CHARACTERISTICS					
Single-family dwelling	61.7%	42.0%	76.1%	80.0%	4.4%
Renter occupied	36.2%	46%	19.4%	19.7%	45.4%
Apartments (20 or more units)	8.8%	23.7%	8.4%	3.2%	82.3%
Median home value	$193,500	$293,000	$460,700	$379,400	$338,900
Affordability					
Mortgage (SMOCAPI)	10.9%	27.9%	32.9%	34.4%	16.1%
Renter (GRAPI)	41.5%	44.3%	47.1%	51.5%	$49.1%
CRIME RATE					
Total per 1,000 residents	2,745[1]	1,871[1]	1,025[2]	1,435[2]	1,026[1]

Sources: 1. https://www.areavibes.com/great+neck+plaza-ny/crime/, undated.
2. https://www.criminaljustice.ny.gov/crimnet/ojsa/indexcrimes/2017-county-index-rates.pdf, 2017.
Notes: All data obtained from U.S. Census American Community Survey 5-Year Estimates (2017) except crime statistics. SMOCAPI=selected monthly owner costs as a percentage of household income; GRAPI=gross rent as a percentage of household income.

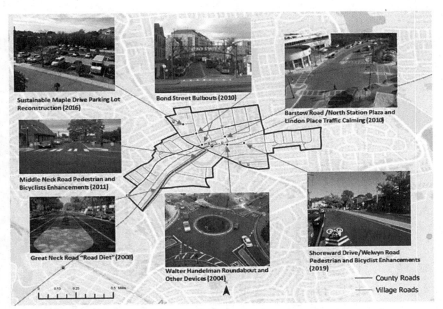

Figure 5.9. Village of Great Neck Plaza road network and Complete Street projects.

for their commute which is far greater than Nassau (17 percent) and Suffolk (6 percent) Counties.

Great Neck Plaza has one LIRR train station and a network of county- and village-owned roads (Figure 5.9). County roads located within the village include South Middle Neck Road and Middle Neck Road (both north-south trending roads) which connect commuters traveling from the LIE/ Northern Parkway north to Kings Point. Three other county roads are Great Neck Road, Cutter Mill Road, and Grace Avenue that connect communities to the west and east. The remaining roads are maintained by the village and mainly connect residential areas to the business district. There are also bus stations located along county roads that are shown on Figure 5.9.

One of the reasons Great Neck Plaza is featured in this book chapter is because they have completed a total of seven sustainable transportation and infrastructure projects and have been a leader in these developments since 2000. These projects are shown on Figure 5.9, and additional information about each project is presented in Table 5.3. In summary, these projects all were funded in large part with federal and state grants. Complete Streets designs included all eight of the design principles presented in Table 5.1 and included the reconstruction of a roundabout, redesign of the roads surrounding the very active LIRR station, putting Great Neck Road on a

Table 5.3. Sustainable transportation projects in Great Neck Plaza

Project	Overview	Total Cost and Funding	Completed
Walter Handelman Roundabout and other Devices	An intersection was redesigned into a roundabout and signs were installed to reduce vehicular speeds, reduce accident frequency and severity, increase pedestrian safety, and reduce the need for police enforcement. It also included overhead speed awareness devices and illuminated advance pedestrian warning devices of midblock crossings.	$640,000 LSSTC (90%) Village (10%)	February 2004
Great Neck Road "Road Diet"	This project renovated a 0.9-mile stretch of road between Middle Neck and Bayview Avenue on Great Neck Road. One lane in each direction was removed, left U-turns were added, wider medians, bulbouts, pedestrian refuges, and safety zones were added. Median streetlighting was relocated to the sidewalk for improved safety and to aid pedestrians which required partnering with the Town of North Hempstead.	$1,191,000 LSSTC (80%) Village (20%)	Summer 2008
Bond Street Bulbouts	Improved two street intersections near the railroad section to make pedestrian crossings to the train station safer by shorting the sidewalk and adding four bulbouts. These curb extensions were constructed to reduce pedestrian crossing distances on Bond Street on the western end from North Plaza to Grace Avenue.	$156,250 LSSTC (80%) Village (20%)	Summer 2010
Barstow Road/ North Station Plaza and Linden Place Traffic Calming	Traffic calming project to improve pedestrian safety by normalizing intersections, adding bulbouts and median islands, shortening crosswalks, adding ADA signs, and providing streetscaping.	$280,000 LSSTC (80%) Village (20%)	September 2010

Project	Description	Funding	Date
Middle Neck Road Pedestrian and Bicyclist Enhancements Project (TEP)	The goals of this project were to improve pedestrian safety on the roads near the train station by redesigning sidewalks and crosswalks, widening medians, and adding better signage and pedestrian ramps. Site furnishings, decorative pavement, trees and planting, public art, and new bicycle racks and overhead canopy to protect bicycles were also added.	$725,000 NYSDOT TEP (80%) Village (20%)	October 2011
Sustainable Maple Drive Parking Lot Reconstruction	Green infrastructure projects that supports many EPA stormwater goals and is an attractive parking facility of 190 parking spaces, with a combination of asphalt and block pavers with native, drought-resistant grasses and other vegetation in a wide natural median. Surrounding the site, the village recently supplemented with specific plant species to attract Monarch butterflies. Mayor's Pledge through the National Wildlife Federation.	$675,000 EFC Green Infrastructure Grant Program (GIGP) (80%) Village (20%)	August 2016
Shoreward Drive/ Welwyn Road Pedestrian and Bicyclists Enhancements (TEP)	The stated project goals were to promote pedestrian, cyclist, and transit use; create safe and attractive street corridors; promote Smart Growth, mixed-land use, and downtown development (including streetscape improvements); increase access for all modes of transportation and the integration of all ages; incorporate the preferences and requirements of the people using the area; increase quality of life; and help reduce the negative effects of motor vehicles on the environment (e.g., pollution, sprawl).	$838,000 NYSDOT TEP (80%) Village (20%)	October 2019

road diet that removed one lane of traffic in either direction, and other traffic calming designs throughout the village (e.g., bulbouts and shortening crosswalks).

These successful transportation projects were accomplished for a variety of reasons most notably leadership from Mayor Jean Celender, who was first elected as mayor in 2000 (Celender 2018). Mayor Celender has lived in the village for 40 years, is a full-time mayor, and is a trained urban planner. When she was elected, she worked closely with the then village engineer to transform the roads using federal grants (Celender 2018). The first project was the Walter Handelman roundabout that was completed in 2004 and was renamed to honor the late village engineer. Successful implementation of these projects has provided the mayor an opportunity to share her expertise, and she was instrumental at providing input that led up to her village adopting a Complete Streets Policy Guide in February 2012 to be consistent with New York State's statewide Complete Streets Policy (Celender 2018).

Under the mayor's administration, three significant policies were enacted all of which dovetail together to guide sustainable development projects in the village. The first policy is a Complete Streets Policy Guide passed in 2012 (Village of Great Neck Plaza 2012). This guide lays out a vision, provides a rationale for the needs and benefits, and provides specific design elements for future road projects. The second policy is a Climate Action Plan (CAP) finalized in February 2016 as part of a Climate Smart Community program (Village of Great Neck Plaza 2016a). The CAP presents a greenhouse gas inventory of both municipal operations and the community and lays out initiatives to address climate. Transportation related initiatives in the CAP focus on "enhancing local and regional connections by increasing the level of bicycling, pedestrian activity, and transit ridership" (Village of Great Neck Plaza 2016a, 18). The third policy, AARP Long Island Action Plan for Age-Friendly Village of Great Neck Plaza, was developed in June 2016 to (1) promote safe, secure and more workable environments for people of all ages and abilities, and (2) allow older residents to enjoy physical independence and advance their mobility (Village of Great Neck Plaza 2016b, 4). Given that a third of the village's population are seniors, any Complete Streets project would need to consider "age-friendly" initiatives laid out in this action plan.

In summary, the Village of Great Neck Plaza has established itself as a leader in the development of sustainable transportation projects through effective leadership, leveraging federal and state grants, and policy adoption and enactment.

We're proud of how we have made infrastructure improvements largely through obtained grants. None of these required bonding and we leveraged village monies to implement these important safety and enhancement projects. They have made the Plaza more walkable, and safer to aid and benefit motorists, pedestrians and bicyclists (people of all ages and abilities) following Complete Streets principles (Celender 2018).

LESSONS LEARNED

Many lessons may be learned by analyzing both Long Island and Great Neck Plaza's transportation systems as summarized below.

Policy can drive action. In New York, there is a mandate to comply with Smart Growth principles and funding for transportation projects must consider Complete Streets design features in order to receive federal and state grants. Both counties have Complete Streets–specific resolutions, and most of the towns have a Complete Streets policy (see Figure 5.8). However, only three villages and the City of Long Beach (out of 96 villages and two cities) have a Complete Streets policy. These villages could easily adopt similar policies as the policy mechanisms (e.g., templated resolutions and action plans) are in place to assist these small villages.

Utilize federal and state grant funding. All seven of Great Neck Plaza's projects were funded mostly from federal and state transportation and green infrastructure grants. Most of these grants are reimbursement grants, and local governments need to have liquid funds available to complete the projects and then seek reimbursement for funds expended (which can be done in partial payments), although many seek bonds to pay for these projects. They must also come up with matching funds (usually 20 percent of the project) which may be difficult for small villages.

Transportation engineering studies and postproject assessments will encourage future projects. In Great Neck Plaza, the mayor recounted how partnering with good engineers is part of the reason for their success (Celender 2018). She said that good engineering design will help produce desired safety results. Numerous postproject engineering studies are presented on the village website that prove that these projects are performing to their desired goals. For example, one study reported that a project resulted in a 64-percent annualized reduction in the total number of injury-reported crashes. Before the project was 55 crashes with 33 after the project (Village of Great Neck Plaza undated). These studies provide efficacy that these projects work and provide rationale and support for future improvements.

Building partnerships and gaining support from stakeholders is critical for success. Throughout these projects, the mayor successfully collaborated with various stakeholders to achieve their visions. Two of the projects required coordination with Nassau County since they own and maintain the roads located within the village boundary. In the development of the AARP Long Island Action Plan for Age-Friendly Village of Great Neck Plaza, a Citizen Advisory Committee (CAC) was created that consists of 17 individuals, a liaison for the AARP and Town of North Hempstead, and the mayor (Village of Great Neck Plaza 2016b). In addition, she holds public meetings for each project and has a transparent public website. The mayor actively presents on their accomplishments at public meetings across Long Island and has presented at Vision Long Island's yearly National Complete Streets meeting (Vision Long Island undated).

CHALLENGES TO IMPLEMENTATION

Even with the grant support, existing policies that promote sustainable transportation, and strong leadership examples, a number of challenges continue to thwart action:

The entire process takes a long time. Long-range planning typically plans 30 or more years in the future. The current regional transportation plan on Long Island lays out a vision for 2045 (New York Metropolitan Transportation Council 2017). The key components of this plan includes developing coordinated development emphasis areas (CDEAS), incorporates Sustainable Communities Initiatives, encourages transit-oriented development, and specifically requires Complete Streets principles and designs for funding. These funds are often distributed through a competitive grant process. It takes time to write, win, and implement a grant project. And there is no guarantee that their project will be funded. Also, administrations can change in the middle of projects.

Local government lacks human and financial capacity. The Village of Great Neck Plaza has been successful because they have a full-time mayor who was formerly trained in urban planning. Most of the village mayors are part time and hold full-time positions outside of civil service, and many villages have a very small staff that lack the specific knowledge to be successful in winning grants. Smaller municipalities also have a lower tax base and simply cannot afford even the matching funds for a grant.

Roads on Long Island are interconnected and require collaboration and support with other local governments. Long Island is a network of state, county, town, village, and privately owned roads. Roads that start in one village can become part of another village for example. This causes issues that require cooperation and commitment from other municipalities which may not be as supportive of the mission. The Village of Great Neck Plaza successfully collaborated with Nassau County and the Town of North Hempstead which shows that these collaborations are possible.

Public perception surrounding mass transit. The perception of the suburbs is that they prefer a lower-density car-dominant lifestyle and it is hard to get buy-in on projects that promote transit-oriented developments, public transit, bicycles, and walking. However, this perception may be changing. Through surveys and stakeholder engagement, the New York Metropolitan Council (NYMTC) reported a need for more reliable, higher-capacity transit service, and improvements to traffic safety. In addition, they indicated that transit-oriented development and initiatives that support sustainable growth are supported by the community (NYMTC 2017). Suburbanites still have concerns over increased traffic congestion; however, they support pedestrian and bicycle projects as well.

CONCLUSION

Long Island has the potential to become the first suburb in America to develop a fully sustainable transportation system. The infrastructure is present with a light rail system, road network, bus lines, and main street development. However, a well-entrenched car culture; lack of public transit everywhere; connectivity, frequency, and capacity problems in the existing system; and lack of financial and human capacity stand in the way of transforming Long Island into a truly sustainable transportation system.

However, recent developments in the Smart Growth and Complete Streets movements show promise in paving the road toward a more sustainable future.

ACKNOWLEDGMENTS

The authors would like to acknowledge Eric Alexander, Director of Vision Long Island; Michelle Gervat, Regional Director at the American Heart Association; and Frank Wefering, Director of Sustainability at Greenman-Pedersen Inc. (GPI) Engineering Firm. These three people are leaders in the Long Island's Complete Streets movement and conceptualized this research project. The author would like to thank Dr. Jessica Holzer, Assistant Professor, Health Administration and Policy at the University of New Haven, and Dr. Sylvia Silberger, Associate Professor of Mathematics at Hofstra University, for participating in meetings, providing general assistance, and participating in the development of the methodological framework. Lastly, this research could not have been accomplished without the financial support from the National Center for Suburban Studies at Hofstra University.

REFERENCES CITED

Advisory Council on Historic Preservation. Undated. "Great Neck Plaza, New York." https://www.achp.gov/preserve-america/community/great-neck-plaza-new-york.

Anderson, Geoff, et al. 2015. "Safer Streets, Stronger Economies: Complete Streets Project Outcomes from Across the United States." *ITE Journal* 85(6): 29–36.

Celender, Jean. 2018. Interview.

Historic Preservation Commission of Great Neck Plaza. Undated. http://www.greatneck-plaza.net/historic/index.php.

Long Island Railroad History. Undated. "Long Island Rail Road History Website." http://lirrhistory.com.

Long Island Traditions and Village of Great Neck Plaza. 2003. *Great Neck Plaza Historic Walking Tour*. http://www.longislandtraditions.org/gnp/gntour.pdf.

National Complete Streets Coalition. Undated. "National Complete Streets Coalition." https://smartgrowthamerica.org/program/national-complete-streets-coalition/.

National Complete Streets Coalition. 2018. "The Ten Elements of a Complete Streets Policy." https://smartgrowthamerica.org/resources/the-ten-elements-of-a-complete-streets-policy/.

National Complete Streets Coalition. 2019. "Complete Streets Policy Inventory." https://smartgrowthamerica.org/program/national-complete-streets-coalition/publications/policy-development/policy-atlas/.

New York Metropolitan Transportation Council. 2017. *Regional Transportation Plan 2045 Maintaining the Vision for a Sustainable Region*. https://www.nymtc.org/Required-

Planning-Products/Regional-Transportation-Plan-RTP/Plan-2045-Maintaining-the-Vision-for-a-Sustainable-Region.

New York State Department of Transportation. Undated. "Complete Streets." https://www.dot.ny.gov/programs/completestreets.

NYSDOT. Undated. "NYSDOT Implementation of the Smart Growth Public Infrastructure Policy Act." https://www.dot.ny.gov/programs/smart-planning/smartgrowth-law.

Sharpin, Anna Bray, Ben Welle, and Nikita Luke. 2017. "What Makes a Complete Street? A Brief Guide." https://thecityfix.com/blog/what-makes-a-complete-street-a-brief-guide-nikita-luke-anna-bray-sharpin-ben-welle/.

Smart Growth America. Undated. "What Is Smart Growth?" https://smartgrowthamerica.org/our-vision/what-is-smart-growth/.

Steg, Linda, and Robert Gifford. 2005. "Sustainable Transportation and Quality of Life." *Journal of Transport Geography* 13: 59–69.

Tri-State Transportation Campaign. 2016. "The Region's Most Dangerous Roads for Walking." http://tstc.org/reports/danger16/index.php.

TSTC. 2017. *Out of Service: 10 Years of Decline in Nassau County's Bus System.* http://www.tstc.org/wp-content/uploads/2017/11/nice-out-of-service.pdf.

United States Green Building Council. Undated. "Leadership in Energy and Environmental Design (LEED) Certification." https://new.usgbc.org/leed.

Village of Great Neck Plaza. Undated. "About Great Neck Plaza and Traffic Calming Initiatives." http://www.greatneckplaza.net/About%20Great%20Neck%20Plaza.html.

Village of Great Neck Plaza. 2012. *Complete Streets Policy Guide.* http://www.greatneckplaza.net/PDF%20files%20and%20forms/Complete%20Streets%20Policy%20Guide%202%201%202012.pdf.

Village of Great Neck Plaza. 2016a. *Village of Great Neck Plaza Climate Action Plan.* http://www.greatneckplaza.net/PDF%20files%20and%20forms/Climate%20Action%20Plan%20Feb.%202016.pdf.

Village of Great Neck Plaza. 2016b. *AARP Long Island Action Plan for Age-Friendly Village of Great Neck Plaza.* http://www.greatneckplaza.net/PDF%20files%20and%20forms/AARP%20Long%20Island%20Age-Friendly%20Action%20Plan%20for%20Village%20of%20Great%20Neck%20Plaza.pdf.

Vision Long Island. Undated. "Complete Streets Coalition Summit." http://visionlongisland.org/complete-streets-summit/.

Zhou, Jiangping. 2012. "Sustainable Transportation in the US: A Review of Proposals, Policies, and Programs Since 2000." *Frontiers of Architectural Research* 1: 150–165.

II

Socioeconomic Sustainable Development in the Suburbs

6

Who Defines Sustainability?

Urban Infill in Colorado Springs, Colorado

JOHN HARNER

For those working to build sustainable development projects, this study highlights what can go wrong and presents warnings to be aware of during the development process. It illustrates what not to do. The case study addresses the process through which an urban infill project adjacent to the University of Colorado at Colorado Springs (UCCS) devolved from a progressive vision of urban sustainability to a generic, placeless landscape. The North Nevada corridor is an urban renewal project intended to stamp a new identity on the blighted district adjacent to the university. Despite the language used by the Urban Renewal Authority to create a walkable, mixed-use, and sustainable project, and in spite of a prime geographic setting, the project devolved into a very traditional, auto-oriented shopping center anchored by big-box stores. A compliant city council and planning commission willingly supported a standard, formulaic design, even issuing dozens of variances to existing regulations to permit the plan. Surveys of students at the university overwhelmingly rejected such an unimaginative project, yet their voices were ignored. The university, while publicly proclaiming to want a pedestrian-oriented destination, a "funky" space that provides a "sense of place" for "student-oriented entertainment" (Burnett 2006; Shockley 2007), actively supported the project with none of these characteristics.

How and why such a placeless landscape was built requires examining how power works to frame the development discourse and shape the development process. The failure of the public participatory process and planning guidelines to affect innovative design reflects a local political culture that permits progrowth coalitions to define "reality" and operate unimpeded in a largely deregulated environment.

CASE STUDY: UNIVERSITY VILLAGE IN COLORADO SPRINGS

The cultural politics of urban growth are concerned with the discursive and material practices in and through which meanings are defined (Laws 1994; McCann 2002; McCann 2004). The ability to fix meaning to commonly used phrases such as *growth, sustainability, diversity,* or *the market* is an exercise in power which sets the terms of public discourse (Ley 1990; Mitchell 2000). Meaning-making discourses are fundamentally intertwined with the place-making politics of local economic development (Laws 1994; McCann 2002; Lees 2004). Actors in the local growth machine construct a discursive frame, or a widely accepted and powerful simplification of the world that "selectively identifies and connects certain elements of everyday life and place in a way that mobilizes and legitimates a particular set of actions or policies while setting other understandings and agendas outside the bounds of consideration" (McCann 2004, 1913). This simplified viewpoint becomes "reality," or what can and cannot be done (Flyvbjerg 1998). Powerful actors produce the reality they want and suppress that for which they have no use.

During public planning meetings, powerful groups can co-opt the language of inclusion while still remaining in control of the process or can simply refuse to participate on equal terms and suppress other moderating or opposing views. Projects become shaped not by the consensus-building outreach but by the elites in charge. The result is a procedural façade (McCann 2001), one that uses the appearance of rationality as a means of legitimation (Flyvbjerg 1998).

Nevada Avenue for decades was the main highway running north-south through Colorado Springs (2015 metropolitan population 678,364). When Interstate 25 was completed in 1961, traffic bypassed the Nevada Avenue corridor and the landscape took on the classic elements of a relict strip: dozens of mom-and-pop motels, automobile-oriented merchants such as gas stations and truck rentals, and a few seedy bars and fast food joints (Harner 2003).

By 2000, the North Nevada Avenue corridor gained the attention of city planners and developers because of its strategic location. This site is about four miles due north of the successful downtown, easily accessible by two exits from the interstate, with the city's main water feature (Monument Creek) and a recreational bike corridor running alongside, and importantly, adjacent to the 540-acre UCCS campus. Both the city and the university hope to build upon the research and development initiatives on campus to create an economic engine around the high-tech and professional industries, the many sports and exercise science centers affiliated with the nearby Olympic Training Center, and the local arts community. As an infill location, the site represented an opportunity to check urban sprawl, serve as the gateway to the downtown, and create a sorely needed housing and commercial environment for students and university employees adjacent to the campus.

Formal attention to the area began in mid-2001, leading to a Reinvestment Plan designed to create "a unique sense of place" (City of Colorado Springs 2004, 4). The plan empowered the Colorado Springs Urban Renewal Authority (CSURA) to eliminate blight "in the public interest" (City of Colorado Springs 2004, 16). Presentations were delivered by UCCS administrators and Regents of the University of Colorado to city council that emphasized the need to redevelop the corridor in order to attract knowledge workers and build an innovation center with "tightly-linked gathering places" (Rutledge 2004, 5). One councilor spoke eloquently of a vision to better integrate UCCS into the city fabric and create lasting character on a unique site (Author Field Notes 2004). The initial phase of the Urban Renewal Plan would be a retail center to drive funding for subsequent development.

The North Nevada Avenue Corridor Urban Renewal Plan (NNACURP) conformed to the city's overall Comprehensive Plan of 2002, specifically with many progressive development standards that embody smart growth principles. These include a heavy emphasis on mixed-land uses, interlinked trails and passageways, respect for existing environmental attributes, multimodal transit access, and developing commercial spaces as activity centers. Together these principles provide a foundation for sustainable growth, development, and design strategies. Promotional pamphlets distributed at public meetings by CSURA showed images of people mingling on streets and sidewalks with shops and cafés alongside bulleted headings "Pedestrian scale," "Memorable experience," "Attractive," "Must be a destination," and "A place people want to spend time." These principles also met with the city's stated Economic Development Corporation (EDC) goals to "work with existing groups to improve the community climate to attract and retain the young professionals that will be an important component of our future workforce" (Economic Development Corporation 2006, 7).

The plan was implemented through Tax Increment Financing (TIF), where bonds provide public funds for infrastructure improvements that facilitate the privately managed economic development and investment. The new commercial outlets are expected to increase property values above those before renewal. The city continues collecting property taxes into its general fund based on property values prior to renewal but increases in property and sales taxes are devoted to paying off the debt on the initial infrastructure improvement. Ideally, the city wants revenues to last far beyond bond timelines. Because of this, the TIF success is tied to project success, long-term viability is a function of project design, and project success or failure can affect city credit evaluations in the future.

Throughout 2005, a private development consortium, Kratt Commercial Properties, began purchasing and consolidating individual parcels into one contiguous unit in anticipation of the urban renewal funding. The developer, aware of the upcoming official designation of the urban renewal area, proactively approached landowners to consolidate parcels. This was done with much communication and coordination between the city and the private developer as the project developed, but before city announcements were made concerning partnerships in the project (Colorado Springs Urban Renewal Authority 2007a; Colorado Springs Urban Renewal Authority 2007b).

On August 30, 2006, a visioning session was held by Kratt employees and CSURA's project manager to present the plan to a number of interested members of the UCCS community. This was the first public presentation of the University Village theme and plan, which consisted of 650,000 square feet of retail space in three large corporate big-box store anchors and numerous smaller spaces, all surrounding a 3,277-space asphalt parking lot—a markedly different option than envisioned in the North Nevada Avenue Corridor Urban Renewal Plan (City of Colorado Springs 2004). The meeting became contentious when university faculty, staff, and students voiced their astonishment. In what would become standard procedure, both the developers and the project manager discounted suggestions for alternative designs.

A subsequent September 13, 2006, public meeting, ostensibly to discuss elements of design, similarly revealed an absence of genuine public input. No substantive elements of the project were under consideration—the public was asked to comment on aesthetics such as paint color for hardware features and landscaping palettes. When participants questioned the very logic of an enormous parking lot–oriented project, that inquiry was summarily dismissed by the project manager. From the very conception, the project design was a fait accompli.

Negative reaction grew on campus. A number of concerned faculty and staff wrote letters to the university chancellor in October and November, 2006, with a list of suggestions for narrower rather than wider streets, greater connectivity between the UCCS property to the east and the commercial center to the west, and many design elements that would generate more public space and smaller-scale buildings, as well as a call to improve the public participation process. The intent was to follow the 2004 NNACURP guidelines to ensure a project that generated a unique sense of place, a destination rather than a park-and-go center.

The ongoing project attracted student attention as well. As part of an undergraduate research project, a UCCS student initiated a survey in February 2007 of student preferences for a university village that showed images of different design options, ranging from standard strip mall to pedestrian-oriented mixed-use projects, and also listed verbal descriptions of different retail environments. With 404 student responses, 95 percent chose images of urban streets and shops with pedestrian traffic (Rockwell 2007). Only 5 percent of students considered designs similar to the one proposed for University Village appropriate. When it became clear that the anchor tenants would be Lowe's (a hardware store) and Costco (a bulk-item grocery and household supply store), neither of which serve the student body, the university shifted their justification for supporting the project by introducing into the discourse the great employment opportunities these businesses would offer to students (Shockley 2007). Here was post-facto rationalization, a new set of criteria used to legitimize the preconceived plan.

A regular occurrence by the major stakeholders throughout the process was to use a narrative with language of sustainability and smart growth yet at the same time support design plans that reflected few of those principles. Such contradictory actions reveal an overtly political tactic. Rather than adhering to a rational process of dialog that evaluates alternative suggestions to come to the best possible plan, growth coalition leaders instead used the sustainability discourse to legitimize their preconceived plan without modification.

At the final planning commission meeting, code violations contained in the plan were defended by the developer, who argued that such infractions were mandatory for success of the project. Variances were subsequently granted, highlighting the ability of the authority figure to define reality even over legal code. Despite discontent on campus, UCCS administration delivered a strong letter of support, claiming that this project was "pedestrian friendly," with "environmentally sustainable designs throughout" (University of Colorado at Colorado Springs 2007). In addition to again using language that misrepresented the actual design, the university speaker delivering the endorsement told the commission that the UCCS Faculty Assembly (among other governance groups) had discussed and approved the plan, when in fact they had not (Author Field Notes 2007; University of Colorado at Colorado Springs 2007). The university defined the "truth"; it had the power make unsubstantiated claims without challenges considered.

On July 24, 2007, the project was presented to city council for final approval. Again, community members voiced concern, particularly because a just-released 2007 study found up to 80 percent retail vacancy rates on a nearby corridor with sites characterized by unneeded parking spaces (City of Colorado Springs 2007a). Many arterial strip mall properties in the Colorado Springs periphery were increasingly abandoned and contributed to a new zone of blight, yet University Village would be another very similar retail center likely to cause increased relocation and further abandonment of retail from those other arterial sites.

All opposition was eloquently countered by the site's developer, Mr. Kratt. During his discussion to city council, he emphasized his plan needed to work *today*, that future agglomerations, or perhaps even building demolition, would only be decided by market forces and were not his immediate responsibility (City of Colorado Springs 2007b). He stated that shoppers demand convenience, and as a builder of shopping centers, that is what he must address. The intent of widening Nevada Avenue and adding traffic lights was to *slow traffic down* and provide quick access for such convenience. Economic reality is such that mom-and-pop stores cannot survive, a point emphasized later by a councilmember who said "the days of locally owned business are gone" (City of Colorado Springs 2007b). Mr. Kratt and UCCS Vice Chancellor Burnett both emphasized the strong support of the university and the long collaboration between them to create a "much friendlier place" that will support research faculty who might, for instance, want to ride their bikes on lunch break. Mr. Burnett again stressed the opportunities for student employment (City of Colorado Springs 2007b). The project passed city council with over 100 code variances that were as yet unresolved. Site grading began in 2008, and the first retail spaces were open for business in late 2009.

University Village is a typical automobile-oriented, suburban-style project (Figure 6.1). Overall, it contains few of the attributes put forth in the Colorado Springs Urban Renewal Guidelines. Seventy-six percent of the property is impervious surface. It has roughly 3,300 parking spaces consolidated in large parking lots, over 1,000 spaces (150 percent) more than required by city code. Buildings are set back from roads and face internal parking lots rather than public spaces or streetscapes. Walls and fences

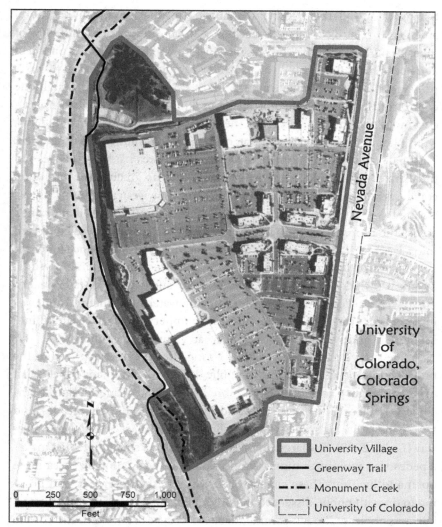

Figure 6.1. University Village footprint. The figure shows big-box stores and enormous, paved parking lots, with no village-like focus and limited connectivity with surrounding land uses. It is designed to ease accessibility by automobiles.

separate retail from adjoining housing and natural water features, and the road width and high-speed traffic of Nevada Avenue are all barriers to pedestrian movement from the UCCS campus. Each building is single story, eliminating the possibility for second-floor professional offices or housing. The physical site was scraped, graded, and filled to overcome the naturally sloping topography, thereby conquering existing place characteristics rather than working with them. Rather than integrating the design into the natural amenities, the project turns its back on a streamside corridor and view of Pikes Peak. The western edge presents 40-foot building backsides on top of a 41-foot terraced retaining wall down to Monument Creek and the city's principal north-south bike trail, the Pikes Peak Greenway. Although one connector trail joins to the north side parking lot, trail users confront an enormous barrier 2,500 feet long. The design was unpopular with students, whose hopes for a service and entertainment environment were ignored. The final plan violated the intent of the city's own streamside ordinances and their mixed-use development ordinances. It has no green space or natural corridors retained for wildlife movement from the western mountains and stream to the open space bluffs to the east. The natural arroyos that drained across the property and provided these wildlife corridors were converted into a large rip-rap settlement structure that fulfills city ordinance for water runoff control, but fails to address strong concerns (and pending lawsuits) by cities downstream about increased flooding and decreased water quality. There were no design elements for capturing water in aesthetically appealing ponds or swales, so that the increase in impervious surface runoff into Monument and Fountain Creeks exacerbates stormwater runoff.

A late surprise came in April 2008, long after the final approval of the project and subsequent decline in public debate. Mr. Kratt announced that his firm intended to implement a more "artistic and hipper design" to include varied rooflines, sandstone and metal on the building facades, and two small gathering spaces with benches and public art near the entranceway (Laden 2008). The larger of these two spaces is a small plaza that makes a pedestrian underpass tunnel connection into the commercial area more open and inviting. These changes, according to the developer, had nothing to do with public dissent to the original plan—they were merely smart business choices. Though commendable, the suggested modifications were largely window dressing to soften the harshness of the enormous parking lot and big boxes (Figure 6.2). More troubling, however, was that Mr. Kratt, in all previous meetings and public events, repeatedly stated the many reasons why such changes could not be made. Whenever challenged on the original design, he had what always seemed to be a sound, economic argument that denied any possibility for deviations. Yet suddenly, without explanation, the developer unilaterally made changes, clearly demonstrating his disproportionate power over public input throughout the entire process. Reality was again redefined by the agent in power.

Figure 6.2. Driveway purely designed for car mobility. The entryway is unsafe for pedestrian walking and unfriendly as a gathering spot.

IMPORTANCE TO THE STATE OF SUSTAINABILITY IN THE UNITED STATES

This project demonstrates how power works to define reality. Rationalization occurs to justify preconceived plans and ideas, which in turn leads to the breakdown between the language of vision statements and general guidelines with what eventually is built. A combination of factors contributed to the project's outcome, but at the highest level is the overarching context of a conservative ideological culture in Colorado Springs that permits (and encourages) such development to occur.

The political culture is distrustful of government, public investments and services, and any regulation over private development. Most people neither want nor expect the government to interfere with the design of land development projects.[1] This neoliberal ideology values the money-making capability of the free market—the economic dimension of development—while downplaying environmental, social, or aesthetic dimensions, even though the public realm is greatly affected when these aspects of development are ignored. Therefore, while planning guidelines may be full of sustainable design principles, discursive practices arbitrate the interpretation of these texts so that city authorities have the power to interpret what phrases like *pedestrian friendly* really mean and how much authority planning guidelines really have. Rules and ordinances can easily be ignored in practice—a clear exercise in power and principal flaw in sustainable planning (Conroy and Berke 2004; Flyvbjerg 1998).

Despite clear problems with long-term viability, Mr. Kratt relied on a standard growth machine narrative. He emphasized new jobs, investment, and tax revenues that would come with the project in the short term. He was able to frame the discourse into his language and vision so that other suggestions simply seem unrealistic. Since he was granted authority as the technical expert and his knowledge was deemed legitimate, the appearance of rationality was maintained in the development process. City council members routinely deferred to his professional judgment, and always what was an "economics issue" overruled any alternative ideas. Future development will achieve the goals of the overall urban renewal guidelines, he repeated, so this initial project need not conform to such rules (City of Colorado Springs 2007b). Yet the last-minute inclusion of design elements long called for, and previously rejected outright, clearly shows that the developer alone fixes the meaning of economic feasibility and what conforms to planning guidelines—what is, in fact, reality.

Because of community disempowerment, the process was not transparent and design plans were presented to other stakeholders as a fait accompli. Student preferences were ignored, an egregious tactic given the project name "University Village." Most importantly, the role of the urban renewal project manager should have been as mediator between developer, city, and other stakeholders in order to reach consensus and compromise (Conroy and Berke 2004). The project manager in this case fully supported the developer's interests and never accommodated alternatives, despite receiving publicly funded compensation from the university.

In the final planning commission meeting, the letter from the chancellor was a glowing endorsement, albeit full of misstatements about sustainable and pedestrian-friendly design.

The university position, unlike that of city council, was not the result of a philosophy that privileges unfettered private development. Rather, the university was very willing to accept almost anything as superior to the existing land use. They had a major concern with the perceived blight, the run-down motels, the correctional rehabilitation center, and the drug monitoring services. Consistently underfunded, the campus planning modus operandi has been to get the most out of any project for the least amount of money—to prioritize cost over vision. The leaders, not well versed in principles of sustainability or smart growth, truly believed this was the best possible project, an example of a successful partnership. It seems that little consideration was given to the long-term sustainability of this project or to the enormous potential for a truly unique design that reflected the site characteristics.

The discourse constructed with language greatly distorting principles of sustainable development suggests that university leaders were actively deceiving constituents to appease the vocal minority while pursuing an alternative agenda. Campus administrators used the language of inclusion to co-opt dissent yet remain in control of the process. Public statements on campus expressed agreement with student and faculty opposition to the plan, yet official correspondence and dealings with the city administrators and the developer consistently and wholeheartedly supported the original developer's design (Kratt 2007; Mayerl 2007; University of Colorado at Colorado Springs 2007). Campus leaders were selective in the information they used in order to control the process.

This case study shows how a "rational" development process is socially constructed to reflect asymmetrical power relations, but it is likely that there are few surprises here. In fact, many who read the timeline of events and see the results will declare this a smashing success, especially given that development policies reflect the local political culture. When a private company, at considerable risk, invests heavily in a project, it can essentially do whatever it wants on its property. This how the process is supposed to work in a neoliberal context. Unfortunately for adherents to this ideology, the Urban Renewal Authority defaulted on the bonds issued to fund the North Nevada corridor in December 2011. By 2016, the project had fallen 50 percent short of forecasts. Attempts to restructure the bond payments were ongoing into 2017, but the public will likely be stuck with more of the tab.

The bond default illustrates that psychological associations people have with place influence a project's success (Relph 1976). The eradication of distinctive places with standardized landscapes is inherently insensitive to the significance of place for human well-being, or to the importance of landscape aesthetics to overall satisfaction and happiness (Florida 2008). University Village is not a place where people like to gather. The social, environmental, and aesthetic dimensions of development were neglected entirely, with a focus solely on the economic. Shopping was reduced to efficiencies, and qualitative considerations about human and environmental sensibilities were ignored. Without a vision of the project in its entirety and consideration for the long-term sustainability, we are left with a sterile place, built to accommodate cars and commerce at the expense of human experience.

Politics around participatory governance and citizenship are conditioned by the urban built environments in which they take place (McCann 2002). The Colorado Springs City Council reflects a laissez-faire political culture, and the built landscape is a direct reflection of this ideology. Despite progressive planning language and an Economic Development Corporation seeking the creative class, an anti-regulatory ideology that privileges short-term economics has produced a landscape that will likely disappoint the innovative entrepreneurs who drive the new economy and place city revenue at risk. For any community with similar progrowth sentiments and without a strong public engagement mandate or enforceable design principles, beware that council members as active agents in the growth machine coalition will likely settle for frictionless mediocrity rather than sustainable design. The result will continue to be homogenized suburban landscapes lacking in identity.

Note

1. Despite the reluctance to regulate private development for the public good, the city has been more than willing to devote public funds toward private projects. For example, the city agreed in 2009 to pay $1.7 million per year for the next 30 years to keep the U.S. Olympic Committee in town with a new headquarters building. Also, in April 2010, the city overruled the Urban Renewal Authority and unanimously approved a new urban renewal project on a tract of undeveloped land on the urban periphery (designated as blight because it had neither streets nor curbs) in order to generate funds to complete a major highway and upscale shopping mall. The willingness to intervene with public funds to subsidize private development is a clear ideological contradiction that again illustrates how "rationality" is created.

Further Materials

University Village website: https://www.uvcshopping.com
Colorado Springs Urban Renewal Authority: https:// www.csura.org
North Nevada Avenue Zoning Overlay Design Guidelines: https://coloradosprings.gov/sites/default/files/design_guidelines_11-9.pdf
Renew North Nevada Avenue Master Plan: https://coloradosprings.gov/sites/default/files/renewnnavemasterplan_draft_03.06.17_combined.pdf
University of Colorado Colorado Springs: https://www.uccs.edu

References Cited

Author Field Notes. 2004. Taken at the city council meeting, Colorado Springs, CO. December 4.

Author Field Notes. 2007. Taken at the Planning Review Commission public review, Colorado Springs, CO, July 5.

Burnett, B. 2006. PowerPoint presentation to the University Club, Colorado Springs, CO, October 23.

City of Colorado Springs. 2004. *North Nevada Avenue Corridor Urban Renewal Plan.* Colorado Springs: Urban Redevelopment Division.

City of Colorado Springs. 2007a. *Academy Boulevard Corridor Conditions Assessment.* Colorado Springs: Comprehensive Planning Division.

City of Colorado Springs. 2007b. Video recording transcript of city council meeting, Colorado Springs City Hall, July 24.

Colorado Springs Urban Renewal Authority. 2007a. Minutes of the special meeting of the Urban Renewal Authority of the City of Colorado Springs, June 7.

Colorado Springs Urban Renewal Authority. 2007b. Minutes of the special meeting of the Urban Renewal Authority of the City of Colorado Springs, July 26.

Conroy, M. M., and P. R. Berke. 2004. "What Makes a Good Sustainable Development Plan? An Analysis of Factors that Influence Principles of Sustainable Development." *Environment and Planning A* 36(8): 1381–1396.

Economic Development Corporation. 2006. *5-Year Strategic Plan.* Colorado Springs.

Florida, R. 2008. *Who's Your City?* New York: Basic Books.

Flyvbjerg, B. 1998. *Rationality and Power: Democracy in Practice.* Chicago: University of Chicago Press.

Harner, J. 2003. "Relict Landscapes of the Colorado Springs Automobile Strips." Lecture presented at the Pioneers Museum, Colorado Springs, CO, October 18.

Kratt, K. 2007. Phone interview with UCCS student Jackie Rockwell, Colorado Springs, CO, March 11.

Laden, R. 2008. "Plans for North Nevada Shopping Center Modernized." *Gazette,* April 17.

Laws, G. 1994. "Urban Policy and Planning as Discursive Practices." *Urban Geography* 15(6): 592–600.

Lees, L. 2004. "Urban Geography: Discourse Analysis and Urban Research." *Progress in Human Geography* 28(1): 101–107.

Ley, D. 1990. "Urban Livability in Context." *Urban Geography* 11(1): 31–35.

Mayerl, J. 2007. Personal communication, Colorado Springs, CO, February 7.

McCann, E. J. 2001. "Collaborative Visioning or Urban Planning as Therapy? The Politics of Public-Private Policy Making." *Professional Geographer* 53(2): 207–218.

McCann, E. J. 2002. "The Cultural Politics of Local Economic Development: Meaning-Making, Place-Making, and the Urban Policy Process." *Geoforum* 33(3): 385–398.

McCann, E. J. 2004. "'Best Places': Interurban Competition, Quality of Life and Popular Media Discourse." *Urban Studies* 41(10): 1909–1929.

Mitchell, D. 2000. *Cultural Geography: A Critical Introduction.* Oxford: Blackwell.

Relph, E. 1976. *Place and Placelessness.* London: Pion.

Rockwell, J. 2007. *A Quick Economic Engine Versus Sustainable Development for Colorado Springs: A Review of the "University Village" Student Survey*. Unpublished undergraduate honor's thesis, Department of Geography and Environmental Studies, University of Colorado at Colorado Springs.

Rutledge, J. 2004. "The Board of Regents' View of the Future of the UCCS Campus." PowerPoint presentation, Colorado Springs, CO, December 14.

Shockley, P. 2007. Comments delivered at the Chancellor's Forum on North Nevada Avenue, Colorado Springs, CO, March 15.

University of Colorado at Colorado Springs. 2007. *An Open Letter to the City of Colorado Springs Planning Commission*. Office of the Chancellor, June 29.

7

Diversity Improves Design

Sustainable Place-Making in a Suburban Tampa Bay Brownfield Neighborhood

E. CHRISTIAN WELLS, GABRIELLE R. LEHIGH,
SARAH COMBS, AND MILES BALLOGG

One Sunday afternoon in June 2017, members of our team met with a small group of residents from the University Area Community, a blighted neighborhood in suburban Tampa where human and environmental health are among the poorest in the state of Florida. We were talking with families about plans to redevelop a seven-acre plot of land in the heart of the neighborhood, with the goal of turning it into a recreational park and community garden. As our presentation concluded, we asked if anyone was interested in participating in the project. No one answered. After a brief silence, one man remarked, "Look, we been sold too many dreams already. It just ain't goona happen." As we would come to learn, this comment was more than just one resident's opinion about the false hopes often generated by well-intentioned university service projects. The statement also turned out to be emblematic of a broader set of experiences and perceptions about living in the community, which included feelings of isolation and disengagement. As one woman stated during the meeting, "I come home from work and stay in my house and don't feel connected. I want to feel belonged, but for me I don't feel like I matter."

In this chapter, we discuss how these and other stories from residents transformed the way in which we approach redevelopment planning, making the process more inclusive and the outcomes more sustainable. We present this case in the context of emerging research on place-making, that is, the transformation of blighted spaces into meaningful places where residents feel safe, healthy, and a sense of belonging, or what scholars have termed "well-being" (Atkinson, Fuller, and Painter 2016). With funding from the U.S. Environmental Protection Agency, we are developing an area-wide plan for the redevelopment of the community's brownfields—properties contaminated, or believed to be contaminated, with pollutants or contaminants (EPA 2017)—that incorporates the social, economic, and political processes of place-making with the greater goal of turning environmental liabilities into community assets.

Reflecting back on our experiences thus far with the area-wide planning process, we discovered that our administrative (e.g., regulatory, municipal, commercial, and so on) understandings and communications about environmental challenges have tended to focus on brownfield communities as physical *spaces*, while local residents often talk about their community in terms of meaningful *places*. This contrast has sometimes resulted in miscommunication or misunderstanding and, more importantly, a lack of enthusiasm on the part of residents to participate in redevelopment efforts, as indicated in our June meeting with residents. Once we realized this divergence in framing redevelopment issues, we were able to pursue a more equitable process, which is providing legitimate opportunities for diverse community members to engage in substantive decision making in planning and development efforts as empowered stakeholders. We argue that this mode of redevelopment is sustainable and resilient over the long term because multiple voices and perspectives are incorporated into the process and residents feel a sense of ownership in the project. In other words, as the title of this chapter asserts: diversity improves design.

Sustainable Place-Making for Equitable Redevelopment

As property values and corporate taxes have increased in inner cities since the 1970s, many industries have relocated to suburban areas, creating environmental challenges for neighborhood planning and development especially when hazardous wastes are involved. Today, nearly half of the U.S. population (approximately 166 million people) lives within three miles of a brownfield site (EPA 2016, v). Examples of brownfields include vacant industrial sites, shuttered factories, abandoned warehouses, and former gas stations or automotive shops, all of which produced hazardous wastes as part of their production or service processes. The resulting contamination of soil and water can have serious, and even fatal, consequences for human health, especially for vulnerable populations such as children and the elderly (Mohai, Pellow, and Roberts 2009). Most of these sites are located in minority and underserved communities, which often lack the capacity or resources to address both the environmental problems as well as the public health impacts that are linked to such problems (Brulle and Pellow 2006). The main question for these communities moving forward is this: How can residents living with suburban brownfields successfully engage in cleanup and redevelopment efforts to create sustainable, healthy, and economically vibrant communities?

We contend that the answer to this question lies in the distinction between space and place (*sensu* Agnew 2011). Social science scholarship has increasingly emphasized the analytical differences between spaces and places (Ingold 2007; Low and Lawrence-Zúñiga 2003; Massey 2005). Spaces are often characterized geographically in terms of their physical attributes and the processes (e.g., flows or movements of people, materials, and wastes) they afford (Castells 2010). Places, on the other hand, are marked by the social, cultural, and symbolic significance to their inhabitants through experiences, memories, and emotions (Lefebvre 1991). In short, places are meaningful spaces that are constituted as palimpsests of historical activities and events on the landscape that produce significance to residents (De Certeau 1984).

From this distinction between spaces and places, social scientists and others have used the term *place-making* to describe the process of shaping the physical and social character of communities at different scales, from neighborhoods to towns and cities (Fleming 2007; Schneekloth and Shibley 1995). Key to this approach is acknowledging a fundamental tension between alternative discourses on spatiality. *Disembedded* perspectives of space reduce places to metrics, which can then be used to manage places and assert control over their form and function (Crewssell 2004). In contrast, *embedded* perspectives of space situate places in cultural and historical context, which allow dynamic forms of contestation and negotiation to emerge (Friedman 2010). Place-making is about shifting the discourse on space from a disembedded perspective to an embedded one in such a way as to democratize decision-making across a community, similar in approach to citizen science initiatives and community-based participatory action research (e.g., Minkler et al. 2008). As a result, the process of place-making often includes a wide array of stakeholders from public, private, governmental, nonprofit, and community sectors with the greater goal of rejuvenating the built and designed environments, improving health and safety, revitalizing businesses, and inspiring people to engage in redevelopment, organizing, and activism (McCann 2002).

Given the potentially large number of stakeholders involved alongside competing perspectives and agendas, it can be especially difficult to re-develop a space into a place in which everyone can develop meaningful "attachments" (Altman and Low 1992; see also Manzo and Devine-Wright 2014). Pierce and colleagues (2011) have suggested the idea of "relational place-making" to describe this friction. Relational place-making is a net-worked process of meaning-making that is constituted by sociospatial rela-tionships among people with contrasting "place-frames" (Martin 2003, 733), or the political positioning and strategic maneuvering of different notions of place (Pierce, Martin, and Murphy 2011, 59). In other words, the process of place-making is inherently social and political as diverse actors compete, negotiate, resist, and accede to different goals of redevelopment and ideals of place. Relational place-making as an analytical construct is intended to draw attention to the social, political, and economic forces that intersect during redevelopment and to highlight the specific conflicts and places they produce (Pierce, Martin, and Murphy 2011, 67). Research suggests that by revealing the forces and resolving the conflicts involved in place-making, redevelopment projects can become sustainable over the long term (e.g., Dorsey 2003; Marsden 2013).

THE SPACE OF THE UNIVERSITY AREA COMMUNITY

The University Area Community (UAC) is an 864-acre underserved resi-dential/commercial suburban neighborhood located in unincorporated Hillsborough County on the northern edge of the City of Tampa. Roughly 75 percent of the approximately 10,500 residents in the neighborhood rep-resent minority groups, primarily Hispanic (39 percent) and Black (33 percent), and 19 percent are U.S. veterans (U.S. Census Bureau 2014). The neighborhood is blighted by vacant and abandoned lots polluted with solid waste, and is surrounded by numerous businesses that produce hazardous wastes (Cardno 2017). In the heart of the community, there is a large brown-field site that is perceived by residents to be contaminated because of its use history and current state of degradation.

The residential area is situated in a densely populated suburban setting and is bounded by established municipalities (Tampa on the south and east, Lutz on the north, and Carrollwood on the west), but is located in unincor-porated Hillsborough County. In short, no municipality has been willing to incorporate this community, making the neighborhood an island in the middle of the Tampa metropolitan area with limited access to social, health, and utility services. Because of its boundedness and close proximity to

Interstate 275 (Florida), the community has been plagued historically with various kinds of brownfields and related redevelopment challenges. For example, Nebraska Avenue, which runs through the western portion of the community, used to serve as the primary thoroughfare to access downtown Tampa before the interstate was sited and, as a result, properties adjacent to this road are contaminated from many years of traffic with automobiles using leaded gasoline (Hafen and Brinkmann 1996). In addition, access to fresh food, health services, and outdoor recreation is extremely curtailed. For these reasons, residents that have been forced to settle here (often from public housing closures in the city) are disproportionately disadvantaged in terms of income, education, critical infrastructures (water, energy, transportation), and social services.

Pejoratively referred to as "suitcase city" by Tampa residents because of the perceived transient nature of the population (89 percent of the homes are rental properties; Brown 1998), this community has been identified as "one of the most economically depressed neighborhoods in Florida" (Smith 2004). Per capita median income has remained steady over the past several years at roughly $12,000, with nearly 60 percent of residents below the 2017 federal poverty level. One hundred percent of K12 students in the UAC are eligible to receive free/reduced lunch (the primary elementary school, Mort Elementary, is a Title I School; Roldan 2016). Many residents struggle with lack of education (32 percent lack a high school education), lack of employment (the unemployment rate is almost twice that of the county), language barriers (51 percent report speaking English "not well" or "not at all" [Diaz and Ortiz 2018]), health problems (e.g., an infant from this neighborhood is twice as likely to die during the first year of life than any other area in Florida [Smith 2004]), and lack of access and mobility to goods and services (nearly 22 percent of residents lack access to a vehicle compared to seven percent in the county). In sum, social and economic challenges in this community, which are integrated with existing brownfields, have constrained efforts to initiate redevelopment of the UAC (Gouldman 1994).

Historically, the UAC had one of the highest crime rates in Hillsborough County, but in the 1990s the area received hundreds of thousands of dollars in federal "weed and seed" grants to weed out violent crime and drug use and then seed social revitalization and economic redevelopment (Greenbaum 1998). In 1996, the District 1 Sheriff's Office was opened on the site of a former crack house in the neighborhood (Lewis 1997). With attention to crime prevention, community residents worked with local law enforcement and crime watch and prevention programs, which reduced overall crime by 31 percent, violent crimes by 82 percent, and sex offenses by 34 percent (Franklin 2004).

In 1998, the University of South Florida Area Community Civic Association and several other community groups united to create the University Area Community Development Corporation, Inc. (University Area CDC), a nonprofit organization dedicated to providing education and job skills training for local residents. Its primary focus is the redevelopment and sustainability of the at-risk areas surrounding the University of South Florida. The University Area CDC heads a coalition of over 175 area businesses, nonprofits, and government agencies that work together on a variety of children's programs, education support, family enrichment, public safety, health services, workforce development, affordable housing, and community-wide volunteer projects in the broader UAC. The University Area CDC has also begun providing basic health screenings (the Florida Department of Health [2011] classifies the UAC as a "Health Professional Shortage Area" and a "Medically Underserved Area") and has partnered with Feeding Tampa Bay, which offers a mobile food pantry (UAC residents miss 1.2 million meals each year; Roldan 2015). Finally, in 2011 several major regional institutions, including the University of South Florida, founded the Tampa Innovation Alliance (now called "Innovation Place") with the primary goal of promoting economic redevelopment in the area (Roldan 2015).

All of these changes have been very positive for the community and indicate significant potential for successful redevelopment. For redevelopment efforts to be sustainable over the long term, it is vitally important to engage in legitimate stakeholder involvement in the planning processes. However, previous planning efforts involved a limited number of area residents through questionnaires and "town hall" style meetings. A needs-based assessment conducted by the University Area CDC in 2015 revealed that nearly 70 percent of respondents had never heard of the Tampa Innovation Alliance or other similar planning efforts. This suggests not only a lack of sustained and meaningful communication regarding planning, but also a lack of community participation.

With decreasing crime and increasing resources for investment and development of the area, the UAC is poised for positive change. At the same time, as community-based organizations, public-private partnerships, and governmental agencies engage in redevelopment planning, it is crucial to integrate these efforts across organizations. For example, the University Area CDC has created a neighborhood revitalization plan focused on health and housing; many local residences were constructed in the 1960s and are suspected to contain asbestos and lead-based paint (Lavely, Blackman, and Mann 1995). The Tampa Innovation Alliance is currently engaged in economic redevelopment planning for the UAC and surrounding area (Benstead 2015). In addition, the Hillsborough County City-County Planning Commission continues to pursue the goals (mostly infrastructure) of its 2001 University Area Community Plan. Notably, all of these plans emphasize assessment metrics and how future development should align with existing policies. While the plans describe "community engagement," this mostly takes the form of public presentations in which residents are informed about what the plans entail.

While existing planning efforts include attention to health, housing, infrastructure, and economic development, none of these plans explicitly recognize the interdependencies between these domains and environmental health. For instance, a recent analysis of federal, state, and local records (EDR 2016) identifies numerous threats to environmental health and safety in the UAC, including 167 facilities that are monitored for regulatory compliance with hazardous waste disposal. These include 14 Florida Super Act Risk sites, mainly petroleum and dry-cleaning facilities investigated for possible contamination of groundwater and drinking wells. Most visible to residents, however, are numerous vacant lots in the neighborhood that community members perceive as contaminated due to the presence of large amounts of solid waste. Finally, brownfield regulatory documents for this area record two large brownfield sites representing former industrial and manufacturing businesses that resulted in the deposition of arsenic (from pressure-treated wood) and lead (from piping and other infrastructure) into subsurface soils adjacent to multifamily residential zones. These brownfield sites occupy the northwest corner and south-central edge of the UAC and have been only partially remediated, although they recently received risk-based closure orders. The presence of these sites along with other forms of pollution in the UAC have created concerns by residents, especially regarding the safety of engaging in outdoor recreation and the ability to participate in community gardening. Overall, there are numerous and overlapping brownfields concerns in the community, and local residents are only variably aware of the health risks imposed by these sources of potential contamination.

The PLACE of the University Area Community

Since the redevelopment planning process began in 2016, we have had the opportunity to speak with numerous residents, formally (with 12 semistructured interviews) and informally (through approximately 600 hours of participant observation), in a wide variety of settings (see Lehigh 2018 for details). In 2015 and again in 2017, the University Area CDC administered a 54-question needs-based assessment using printed, online, and walking surveys in the community, with a total of 875 respondents (60 percent female, 40 percent male, 48 percent ages 25–44). The survey covered issues related to education, employment, transportation, housing, health and safety, and other topics (Diaz and Ortiz 2018). From these forms of communication and interaction, we have heard what it is like to live in the UAC on a daily basis and what residents imagine for future redevelopment.

In terms of education, many residents report working on their GEDs (82 percent have earned or are earning their GED). Approximately 80 percent of people we spoke with have no college degree or even envision college as a possibility, largely due to perceptions of cost but also lack of transportation and childcare. Still, even those with a GED lament not being able to leverage their education to obtain a good-paying job. One man we spoke with said, "I have a GED but I can't help my nine-year-old with his math homework. Why not? What's the use of education then? The kids get aggravated and I get frustrated. It makes me very sad." Another woman we spoke with, who has an elementary level education, finds it difficult to keep her 16-year-old son in school. Education is often seen as "a way out" for children, but many residents struggle to keep their children in school and feel uncertain that the education their children receive will be sufficient for them to gain employment. We also learned that "a lot of kids have kids," as one resident stated, and because the parents are so young, the children end up living with their grandparents and moving in and out of the community throughout the year. The result, as a local elementary school principal observed, is that "every year I get a whole new set of kids, so it's really hard to invest in one cohort and keep that momentum going."

Unemployment in the community is very high. Seventy percent of survey respondents reported being unemployed or underemployed. We heard many reasons for this, including lack of education, lack of opportunity, lack of transportation, felony history, and immigration status (a significant proportion of residents do not have legal residency status). Health-related issues also preclude many from working. One man we spoke with told us that he has been on disability for 30 years and is not allowed to work or else "the government will take away my services. Sometimes I do dishes or deliver pizza, but I keep it a secret or I'll lose my benefits. I'm better off staying under the table." Many employment problems are also linked to transportation. A woman who has a job at a shipping warehouse in a nearby city depends on public transportation to make it to work. She told us, "It's only 10 miles away, but it takes me two hours to get there. And if the bus is late or I miss it, I get in real trouble. I used to work at McDonald's but I lost it because I was late too many times."

Lack of access to transportation is one of the most significant challenges for people living in the UAC. Our survey indicates that approximately 30 percent of people in this community do not own a vehicle and nearly 60 percent depend on public transportation. One woman told us that "the biggest problem is that there's no buses when I get off work late at night, so it's really hard for me to get home." Some residents also feel that the buses are not clean or safe. As one man stated, "The buses are filthy. People smoke on the buses, tobacco and marijuana. I complained, and now they don't allow me on the bus anymore." Unfortunately, the local bus line that runs through the community was recently cut, creating major transportation challenges for education, employment, and access to health services. Several residents reported not being informed about the changes until it was too late and they missed work.

Housing is one of the most critical challenges in the UAC, especially because the vast majority (93 percent) of residents rent apartments, townhomes, or mobile homes. "Bad credit" and "unstable income" were the top two explanations for lack of home ownership. Almost everyone we spoke with complained of having a negative experience with their housing situation. One woman said she has no heat or air-conditioning and, "when I complained to code enforcement, they came and told the landlord, and then I got kicked out of my apartment [for complaining]." We learned that many apartment complexes are owned by out-of-state corporations that often change the management of the facility, making it difficult for renters to have their problems addressed. As one resident stated, "The slum lords treat you like you ain't nothing. Due to the fact that you are less fortunate." Housing troubles also can have serious consequences for health. Many residents reported problems with mold, for example. One man told us, "My friend's apartment has a lot of black mold. Her children are sick all the time." Other residents have water that is so contaminated they use bottled water to bathe and cook. As one woman remarked, "The property we live at, I feel like it is contaminated. We can't drink our water there, you end up going to the hospital." Another woman commented that there is "rust in my water, when you turn it on, red and brown water comes out. I'm scared to say anything to the landlord or I'll get kicked out, and I have nowhere else to go."

Finally, residents reported being concerned about several health and safety issues throughout the neighborhood. Approximately 65 percent of survey respondents noted constant crime in their neighborhood. The most predominant crimes cited were drugs, burglaries, vandalism, domestic violence, gang activity, child abuse, and human trafficking. As one resident put it, "I would love to take my kids to the park, but they are too scared of drugs and getting hit by cars." Another resident told us, "There are abandoned lots where squatters and homeless stay and you [see] people fighting in them." In addition to problems with drugs and crime, residents noted the lack of infrastructure. One man stated, "We need sidewalks. It's not safe to walk in the street all the time. There's nowhere to walk. You see a lot of women pushing strollers. They can't do that in the dirt." Another man replied, "Where I live the road is dirt. We don't have paved roads even." The lack of sidewalks and traffic calming measures in roadways has had serious consequences. In 2014, a six-year-old boy was struck and killed by a speeding car as he was playing on his scooter along the side of the road. There is also no contiguous wastewater infrastructure in the community, which results in long-lasting floods during torrential rains. Many residents complain not only of stormwater flooding but also the problems associated with standing water. As one man reported, "My apartment has floods and now has mold. Horrible flooding. I complain, but nothing ever happens." Another man said, "All of 20th Street floods when it rains. And it just sits there. And then the mosquitos come."

When asked about the possibilities for redevelopment of the park area, residents reported wanting a place where they could exercise and socialize. One woman new to the U.S. said, "I want to lose weight. I gained a lot of weight coming to the U.S., eating processed foods. When you get off work, you're tired. Fast food is easy and the dollar menu fits my budget. But how about a garden? I used to garden in my country and eat vegetables all the time." Another man contributed, "We also need a place to do sports activities, ride bikes, and throw balls. There's no place to do that now." A woman added, "I stay in my house all the time. I don't feel connected. I cannot trust my neighbors. We need a place for social gatherings, to get to know your neighbors better." Summarizing these sentiments, one resident remarked,

"I didn't have a chance as a child to be a child. We need spaces for them to play."

We often conclude our interviews and focus groups by asking what residents perceive are the strengths in the community. One man summed up his views for us, which provided a flashpoint for our planning process. He said, "Engaged people are the strengths, talking up, speaking out." Another man said it is everyone's responsibility to be engaged, "I did my homework, now it's time for footwork." From these statements and all the stories we heard, we realized that the community is more than a space to move around in. It is a place that holds a great deal of meaning for the residents and, while it has many challenges, it also has residents who want to do something about them.

SPACE AND PLACE IN PLANNING AND REDEVELOPMENT

The University Area CDC needs-based assessment concluded that the primary needs identified by residents concern human and environmental health. While we had empirical data to support this claim, we needed to hear from community members to help us bridge the space/place divide. Residents specifically wish to have access to a recreational park in the neighborhood (there currently are no places for children or others to play outside of small apartment complex grounds), especially one in which they can develop a community garden and gain access to healthy foods (the primary grocery store serving this area recently closed, leaving many residents with very limited access to fresh foods). In early 2017, we received a grant from the U.S. Environmental Protection Agency to develop a brownfields area-wide plan for the community. The area-wide planning project allows us to pursue a holistic planning process for the entire neighborhood that takes into account the integrated nature of the community's challenges as well as residents' sense of place.

The area-wide planning process began before we received funding from the EPA. Throughout 2016, project partners met to discuss the redevelopment planning effort. These partners included community residents, the University of South Florida (located adjacent to the UAC), the University Area Community Development Corporation, Mort Elementary School, the Environmental Protection Commission of Hillsborough County, the Florida Department of Health, Hillsborough County Economic Development, Hillsborough County City-County Planning Commission, and three private consulting and social marketing firms. The group decided that one way to respond to residents' concerns while building on the community's strengths is to approach the planning process with the theme of turning brownfields into "healthfields" (Ballogg 2015), which would provide opportunities for outdoor recreation, community gardens and other healthy food options, and access to health care as well as improved housing. The catalyst site selected for our project was a seven-acre parcel with a pond that was acquired by the University Area CDC from Hillsborough County. Once a formal park was established, the place could be used for programming and capacity building in the community.

Our first priority for the area-wide planning project was to collect data on all the facets of community life that could inform redevelopment. These rapid assessments included a brownfields environmental inventory (overseen by Cardno, a local environmental consulting firm), a social impact assessment (facilitated by BGW Associates, a local environmental justice consultancy in the UAC), a health impact assessment (facilitated by the Florida Department of Health), and an economic market assessment (facilitated by Hillsborough County Economic Development). Students from USF also participated in these studies. For example, after the brownfields inventory was conducted, students from a graduate seminar on environmental justice conducted ground-truthing (Sadd et al. 2013) of all the listed sites, documenting and updating the current status of each. University students also participated in the social impact assessment, providing historical and contemporary land use data for the analysis.

To integrate these various strands of data, we worked with the University of South Florida's Digital Heritage and Humanities Collections to create a searchable online archive of all the data we collected. This archive allows the public to search the environmental inventory as well as census data and information from the county planning commission and the metropolitan planning organization among other datasets. For example, the archive includes historical aerial imagery of the area as well as a three-dimensional model of the park. The archive also integrates the social, economic, and health assessment data into a GIS format so that users can see how different datasets overlap with or diverge from one another. Phase two of the project is ongoing and involves additional community collaboration and feedback using the online archive as a planning tool to consider alternative redevelopment scenarios.

Community outreach and collaboration has been key to the project's success thus far, and has involved community meetings and workshops at the park, the needs-based assessment survey, monthly meetings and capacity building with a community-based leadership group, and many other activities—most of which are coordinated by the University Area CDC. We also worked with a local communications and marketing firm (Vistra) to develop a website and various social media platforms to share information and receive feedback on the planning process. Perhaps one of the more unique approaches that have been taken to community engagement has been a design charrette with local children (Sutton and Kemp 2002), specifically aimed at soliciting input from them (and their parents) on the design of a playground for the park.

While the planning process for the community continues, the University Area CDC is beginning redevelopment efforts at the park, now named "Harvest Hope Park." For example, the University Area CDC has constructed a vegetable garden in 33 raised beds on the property. A small building onsite has been redesigned to serve as a model kitchen and is used to teach area residents how to prepare healthy meals using produce from the garden (McKenzie 2015). Working with community partners, the University Area CDC recently built a KaBOOM! playground for the park and is in the process of developing a multipurpose sports field on the property, walking trails around the pond, and other amenities. The University Area CDC also partnered with a local arts foundation to establish "art in the park," which is viewed as a cornerstone for creative place-making (Fleming 2007). Current projects include murals and sculptures designed, created, and installed by community residents to engender a sense of place (Figure 7.1).

The development of this park has high potential to spur redevelopment in surrounding properties (residences and vacant lots), which in turn will play a key role in further redevelopments to surrounding businesses, local schools, and other commercial and service-oriented properties. The University Area CDC, for instance, is engaged in a land-banking effort to acquire properties adjacent to the park, and is working with Habitat for Humanity to construct affordable housing on these lots for single-parent families. All of these changes can positively impact job creation for the area and therefore address concerns regarding unemployment and poverty. In many ways, area residents view the social, economic, and environmental changes to Harvest Hope Park as the key to redevelopment for the entire community.

Figure 7.1. Art in the Park, at Harvest Hope Park in the University Area Community, Tampa, Florida. "Love" (*left*) and "Family" (*right*) are sculptures created and installed by University Area residents with the support of the Gobioff Foundation, a nonprofit organization dedicated to creative place-making in Tampa.

Challenges and Opportunities for Sustainable Place-Making

In some ways, the UAC has been for many years what geographer Ruth Gilmore (2008) calls a "forgotten place." Forgotten places are "'between' or marginal places, not quite urban and not quite rural," where residents are "exhausted by the daily violence of environmental degradation, racism, underemployment, overwork, shrinking social wages, and the disappearance of whole ways of life *and* those who lived them" (Gilmore 2008, 32, emphasis original). But there is finally change on the horizon for the UAC. Redevelopment planning that puts residents first and gives them a voice and a vote in what will happen has been important to this change. Such recentering of redevelopment planning is vital because—as one of our environmental justice colleagues once remarked—"if you're not at the table, then you're probably on the menu."

Still, for as many successes we have seen, there have been challenges and setbacks. Reflecting back on our early efforts, we realized that much of the planning that has taken place already, and indeed some of our own planning efforts, treated the community as a space, where plans were focused on the materiality of the space: form, function, services, and infrastructure— all carefully aligned to comply with external policies and "best practices." Presentations to community members by our team and by other planning groups historically did not elicit much interest or participation by residents. But when we began to have open conversations with residents about their concerns and desires for their neighborhood, we realized that place was a more important conception for redevelopment than space. By focusing on the meanings that people imbue spaces with, we were able to develop a more complex understanding of the "differentially distributed meanings, experiences, local knowledge and individual, as well as collective, understandings of places, spatial relations and representations" (Low 2017, 68). Shifting the discourse from space to place in public meetings and other gatherings has allowed residents to imagine what the park could become, instead of focusing narrowly on what is wrong with the space.

To neutralize the tensions between space and place, we needed to overcome two major challenges. First, we needed to change communication styles. Most of us on the planning team are outsiders—we do not live in the community. As such, residents' stories about their use of the spaces and the meaningfulness of the places became increasingly important turning points for community engagement. At the same time, space metrics for redevelopment were important for communicating with other outsiders about policies, regulations, funding, and other political-economic dynamics of redevelopment. Changing our communication styles to accommodate both audiences helped us navigate the planning process more effectively.

The second hurdle for bridging the space/place divide was translation. The members of the planning team and the residents of the community did not share the same worldviews. Intersecting issues of race, class, and gender differently positioned each one of us with separate but overlapping interests, concerns, outlooks, and opinions (see Pulido and Peña 1998). In addition, some team members came from different disciplinary backgrounds—social science, engineering, environmental science, public health, business and industry, and so on. Translating ideas between disciplines, while also being sensitive about our own positionality, was not always easy but was essential for developing a broader understanding of each other's worldviews, values, and beliefs.

In this way, our planning project has tried to give attention to both space and place. We used established social and environmental science methods to collect evidence on human and environmental health, quantified and assessed the evidence, and then used these assessments to build models for better understanding threats to public health and the environment. At the same time, interviews with residents were used to understand the nuances of community engagement and stakeholder involvement at all levels of the project as well as inform us about challenges and opportunities associated with the redevelopment of the park.

We also have attempted to build capacity for community residents to address challenges on their own terms. Environmental mapping with students and residents working together was an important precursor to formal environmental impact assessments, and allowed us to identify locations in the community that residents perceive to be polluted, contaminated, or otherwise difficult to redevelop. This activity also allowed us to examine extant and emerging environmental justice issues in the community. The social and health impact assessments (making use of ethnography, interviews, and archival/public documents research) allowed us to anticipate impacts on the community with regard to the redevelopment projects and determine how best to proceed with public consultation and the integration of technical expertise in the planning process. Finally, the economic market analysis informs us about community needs for businesses and services, how to prioritize those needs, and where such services might best be located for community members. In this way, sustainable place-making supports evidence-informed decision making for positive policy development to improve the welfare of community residents.

ACKNOWLEDGMENTS

This research was supported by a grant from the United States Environmental Protection Agency (66.814 Brownfields Training—Research and Technical Assistance Grants and Cooperative Agreements), Cooperative Agreement No. D54917. The work was performed with the permission of the Institutional Review Board of the University of South Florida, IRB Study No. Pro00028423. We gratefully acknowledge the advice and support of Nestor Ortiz and Diana Diaz from the University Area Community Development Corporation; Tanya Camacho from Cardno; Beverly Ward from BGW Associates; Allison Nguyen and Ayesha Johnson from the Florida Department of Health; Rebecca Hessinger from Hillsborough County Economic Development; Allison Amram from the Environmental Protection Commission of Hillsborough County; Ed Stillo and Julie Capobianco from Vistra; and Mark Hafen, Lori Collins, Travis Doering, and Richard McKenzie from the University of South Florida.

REFERENCES CITED

Agnew, John A. 2011. "Space and Place." In *The SAGE Handbook of Geographical Knowledge*, edited by J. A. Agnew and D. N. Livingstone, 316–330. London: SAGE.

Altman, Irwin, and Setha M. Low. 1992. *Place Attachment*. New York: Plenum.

Atkinson, Sarah, Sara Fuller, and Joe Painter, eds. 2016. *Wellbeing and Place*. London: Routledge.

Ballogg, Miles. 2015. "Brownfields to Healthfields: Presentation for the Federal Interagency Working Group on Environmental Justice." Last accessed September 21, 2019. https://www.epa.gov/environmentaljustice/brownfields-healthfields-florida-healthfields-successes.

Benstead, Justine. 2015. "Tampa Innovation Alliance: Creating Plans, Partnerships and Sense of Place." *83degrees*, January 13. Last accessed September 21, 2019. http://www.83degreesmedia.com/features/innovationalliance011314.aspx.

Brown, Glenn R. 1998. *An Examination of Homeless Men, Day Labor, and the University West Community*. MA thesis, University of South Florida, Tampa.

Brulle, Robert J., and David N. Pellow. 2006. "Environmental Justice: Human Health and Environmental Inequalities." *Annual Review of Public Health* 27: 103–124.

Cardno. 2017. *University Area Community Brownfields Environmental Inventory*. Report submitted to the University Area Community Development Corporation, Hillsborough County, Florida.

Castells, Manuel. 2010. *The Rise of the Network Society*, 2nd ed. Oxford: Blackwell.

Cresswell, Tim. 2004. *Place: A Short Introduction*. Malden, MA: Blackwell.

De Certeau, Michel. 1984. *The Practice of Everyday Life*. Berkeley: University of California Press.

Diaz, Diana, and Nestor Ortiz. 2018. *2017 University Area Community Survey: Summary Report*. Tampa, FL: University Area Community Development Corporation.

Dorsey, Joseph W. 2003. "Brownfields and Greenfields: The Intersection of Sustainable Development and Environmental Stewardship." *Environmental Practice* 5(1): 69–76.

EDR. 2016. *EDR DataMap Area Study: University Pilot Area*. Environmental Data Resources Inc., Shelton, CT.

EPA. 2016. *Fiscal Year 2017: Justification of Appropriation Estimates for the Committee on Appropriations*. U.S. Environmental Protection Agency, Washington, D.C.

EPA. 2017. "Overview of the Brownfields Program." Last accessed September 21, 2019. https://www.epa.gov/brownfields/overview-brownfields-program.

Fleming, Ronald L. 2007. *The Art of Placemaking: Interpreting Community Through Public Art and Urban Design*. London: Merrell Publishers.

Florida Department of Health. 2011. *Hillsborough County Health Department 2010/2011 Community Health Profile*. Hillsborough County Health, Florida Department of Health.

Franklin, Marcus. 2004. "Where Progress Hopes to Settle." *St. Petersburg Times*, May 13. Last accessed December 19, 2017. https://www.newsbank.com.

Friedman, John. 2010. "Place and Place-Making in Cities: A Global Perspective." *Planning Theory & Practice* 11(2): 149–65.

Gilmore, Ruth W. 2008. "Forgotten Places and the Seeds of Grassroots Planning." In *Engaging Contradictions: Theory, Politics, and Methods of Activist Scholarship*, edited by C. R. Hale, 31–61. Berkeley: University of California Press.

Gouldman, Steven E. 1994. *Applied Anthropology, Participatory Planning and the USF Area Neighborhood Planning Study*. MA thesis, University of South Florida, Tampa.

Greenbaum, Susan D. 1998. "The Role of Ethnography in Linkage with Communities: Identifying and Assessing Neighborhoods' Needs and Strengths." In *Promoting Cultural Competence in Children's Mental Health Services*, edited by M. Hernandez and M. R. Isaacs, 119–132. Baltimore, MD: Paul Brookes Publishing.

Hafen, Mark R., and Robert Brinkmann. 1996. "Analysis of Lead in Soils Adjacent to an Interstate Highway in Tampa, Florida." *Journal of Environmental Geochemistry and Health* 18: 171–179.

Ingold, Tim. 2007. "Earth, Sky, Wind, and Weather." *Journal of the Royal Anthropological Institute* 13: 19–38.

Lavely, Lynn, Joe Blackman, and Karen Mann. 1995. "University West, USF Area, A Demographic and Socioeconomic Profile, A Discussion of Positive Elements in the Community, A Needs Assessment." Institute for At-Risk Infants, Children, and Youth, and Their Families, College of Education, University of South Florida, Tampa.

Lefebvre, Henri. 1991. *The Production of Space*. Oxford: Blackwell.

Lehigh, Gabrielle R. 2018. *Capacity Building, Environmental Justice, and Brownfields Redevelopment: A Case Study of Harvest Hope Park, Tampa Bay, FL*. MA thesis, University of South Florida, Tampa.

Lewis, Harold W. 1997. *An Anthropological Analysis of the Development of a High Crime Area Around the University of South Florida*. PhD dissertation, University of South Florida, Tampa.

Low, Setha M. 2017. *Spatializing Culture: The Ethnography of Space and Place*. London: Routledge.

Low, Setha M., and Denise Lawrence-Zúñiga, eds. 2003. *Anthropology of Space and Place: Locating Culture*. Oxford: Blackwell.

Manzo, Lynne C., and Patrick Devine-Wright, eds. 2014. *Place Attachment: Advances in Theory, Methods and Applications*. New York: Routledge.

Marsden, Terry. 2013. "Sustainable Place-Making for Sustainability Science: The Contested Case of Agri-food and Urban-Rural Relations." *Sustainability Science* 8: 213–226.

Martin, Deborah G. 2003. "'Place-Framing' as Place-Making: Constituting a Neighborhood for Organizing and Activism." *Annals of the Association of American Geographers* 93(3): 730–750.

Massey, Doreen. 2005. *For Space*. Thousand Oaks, CA: Sage.

McCann, Eugene J. 2002. "The Cultural Politics of Local Economic Development: Meaning-Making, Place-Making, and the Urban Policy Process." *Geoforum* 33: 385–398.

McKenzie, Joyce. 2015. "UACDC Launches Kitchen to Harvest Healthy Eating Habits." *Tampa Tribune*, March 25. Last accessed December 19, 2017. https://tampabay.newspapers.com/image/343278383/?terms=%22UACDC%2BLaunches%2BKitchen%2Bto%2BHarvest%2BHealthy%2BEating%2BHabits%22%22.

Minkler, Meredith, Victoria Breckwich Vasquez, Mansoureh Tajik, and Dana Petersen. 2008. "Promoting Environmental Justice Through Community-Based Participatory

Research: The Role of Community and Partnership Capacity." *Health, Education, and Behavior* 35(1): 119–137.

Mohai, Paul, David Pellow, and J. Timmons Roberts. 2009. "Environmental Justice." *Annual Review of Environment and Resources* 34: 405–430.

Pierce, Joseph, Deborah G. Martin, and James T. Murphy. 2011. "Relational Place-Making: The Networked Politics of Place." *Transactions of the Institute of British Geographers* 36: 54–70.

Pulido, Laura, and Devon Peña. 1998. "Environmentalism and Positionality: The Early Pesticide Campaign of the United Farm Workers' Organizing Committee, 1965–71." *Race, Gender & Class* 6(1): 33–50.

Roldan, Roberto. 2015. "University Area Hopes Innovation Alliance Leads to Greater Promise." *Tampa Bay Times*, December 3. Last accessed September 21, 2019. http://www.tampabay.com/news/business/economicdevelopment/university-area-hopes-innovation-alliance-leads-to-greater-promise/2256300.

Roldan, Roberto. 2016. "Suitcase City: Tampa Neighborhood Commits to Help Troubled School." *WUSF*, May 16. Last accessed September 21, 2019. http://wusfnews.wusf.usf.edu/post/tampa-neighborhood-commits-help-troubled-school.

Sadd, James, Rachel Morello-Frosch, Manuel Pastor, Martha Matsuoka, Michele Prichard, and Vanessa Carter. 2013. "The Truth, the Whole Truth, and Nothing but the Ground-Truth: Methods to Advance Environmental Justice and Researcher-Community Partnerships." *Health Education and Behavior* 41(3): 281–290.

Schneekloth, Lynda. H., and Robert. G. Shibley. 1995. *Placemaking: The Art and Practice of Building Communities*. New York: Wiley.

Smith, Wes. 2004. "Hit-and-Run Wounds Neighborhood." *Orlando Sentinel*, April 7. Last accessed September 21, 2019. http://articles.orlandosentinel.com/2004-04-07/news/0404070262_1_step-forward-teacher-wilkins.

Sutton, Sharon E., and Susan P. Kemp. 2002. "Children as Partners in Neighborhood Placemaking: Lessons from Intergenerational Design Charrettes." *Journal of Environmental Psychology* 22: 171–189.

U.S. Census Bureau. 2014. American Community Survey. Last accessed September 21, 2019. https://factfinder.census.gov/faces/nav/jsf/pages/index.xhtml.

8

Sustainable Economic Development

The Maker Movement in Macon, Georgia

SUSAN M. OPP

Over the past thirty years, policymakers across the globe have called for the pursuit of sustainable development or, more specifically, "development that meets the needs of the present without compromising the ability of future generations to meet their needs" (Brundtland 1987). Even with the widespread recognition of the importance of sustainable development the pursuit of this goal has proven to be a difficult challenge for most American cities. In the face of deindustrialization, a changing global economy, and ongoing recessionary pressures many cities struggle to achieve a balance between their immediate economic development goals and their longer term sustainability concerns (Opp, Osgood, and Rugeley 2014). In light of these challenges, researchers have moved to examine the intersection of economic development policy and sustainability policy to see what relationships may exist and to perhaps offer some policy guidance for cities (Deslatte and Stokan 2017; Osgood, Opp, and Demasters 2017). Thus far the conclusions of this research do not show a widespread integration of economic development and sustainability policy in American cities. Furthermore, this research has provided some evidence that competition for economic development as well as economic pressures can lead even the most sustainable city to return to less sustainable approaches to economic development (Opp, Osgood, and Rugeley 2014; Osgood, Opp, and Bernotsky 2012).

The ongoing struggle to balance a city's immediate economic needs with longer term sustainability goals has prompted many organizations and researchers to search for new approaches to achieving this balance. As one example, in the last few years the United States Environmental Protection Agency (EPA) developed a new technical assistance program directed, at least in part, at using local food systems as a tool for sustainable economic revitalization (EPA 2017). Another recent example of an emerging sustainable development approach is that of the maker movement and community makerspaces. Although relatively new to the sustainable economic development discussions, the maker movement is rapidly emerging as a potentially transformative approach for American cities seeking to find a balance between economic concerns and sustainability goals. In recent years two national-scale organizations, Recast City and the National League of Cities (NLC), have both identified the maker movement as a potential tool for sustainable development.

THE MAKER MOVEMENT

The maker movement generally refers to an eclectic and diverse set of individuals associated with a do-it-yourself (DIY) approach to creating new and unique products. Exactly what product is being created varies widely, and this movement does not take one definition or form across the United States. Even with the diversity found in this topic there are some common themes that are instructive to highlight. First, the maker movement is often associated with an end goal of providing an opportunity for individuals to engage in small-scale manufacturing that may lead to individual and/ or community economic benefits. Second, the maker movement is usually meant to be a collaborative endeavor whereby the spaces used for facilitating the DIY work, known as makerspaces, are designed to be shared. In addition to the actual sharing of the physical space, makerspaces are also meant to facilitate the sharing of ideas among participants. Finally, when a local government supports a local maker movement, workforce development and/or entrepreneurship often becomes an added feature to help support local economic development goals. Sometimes the workforce development dimensions are expanded to include socially focused programming geared toward encouraging women, minority, or veteran-owned business creation within a community.

The maker movement can help facilitate sustainability in a community across all of the three accepted dimensions of sustainability: environmental, economic, and social. From an environmental perspective, the makerspaces associated with this movement are often placed into downtown districts—many of which are declining and/or vacant. Rather than investing into development on green spaces, makerspaces are able to reuse buildings, direct foot traffic into downtowns or main streets, and contribute to environmental sustainability goals by encouraging reuse and redevelopment as opposed to new, environmentally detrimental development. From an economic perspective, providing a pathway to economic development that encourages small-scale manufacturing and entrepreneurship allows for economic growth that isn't predicated on traditional top-down incentive-based policy approaches. This bottom-up, individual-scale approach provides a pathway for upward mobility and investment into individuals within a community and perhaps contributes to longer-term economic sustainability. Finally, from a social perspective, these makerspaces and the encouragement of the maker movement allows communities to invest into improving unemployment rates, poverty levels, and inequalities through the attention to individual-level participants rather than to attracting investment from outside the community. The participants in the maker movement can begin to innovate, work for themselves, and provide a more socially sustainability pathway to improving the quality of life in the community.

THE GROWTH OF THE MAKER MOVEMENT

The maker movement in the United States is most often traced to the January 2005 publication of Make Magazine where DIY projects were profiled for the first time at a national scale. Later that same year the website Etsy emerged as a platform where individuals could sell their DIY projects and, in essence, have a path to becoming an entrepreneur without needing access to a lot of capital. By April of 2006, this phenomenon had significantly expanded, and the first annual event dedicated to DIY-oriented projects, the Maker Faire, was held in San Francisco. Eleven years later, in 2017, two major and over 200 smaller-scale Maker Faires were held each year attracting more than 200,000 participants across the country (Makerfaire.com). By some estimates, more than 2.3 million people have attended maker-focused events across the world since this movement began (Kalil 2017). Alongside the fragmented growth of the maker movement through these events, this topic gained the attention of many public-sector entities when, in May of 2014, President Obama announced the first Mayors' Maker Challenge (NLC 2016). Today, the maker movement has become a widely accepted potential tool for local sustainable development goals. However, even with the rapid growth and acceptance of this movement, the participation of local governments in this area is quite varied. Furthermore, no singular source of information is available that documents either the level of engagement in the maker movement or the benefits, if any, that are actually being derived from these activities. From the limited information that is available on this topic, it is clear that some local governments aggressively encourage the growth of the maker movement within their cities. However, in other cases, there is evidence that the maker movement is nonprofit-dominated and the local government is an idle bystander.

To be sure, much work remains to be done to fully understand the diversity in approaches and the overall economic benefits obtained from makerspaces. However, even with the limited information available on this topic, it is an important area to consider for local sustainable development purposes. The focus on shared space, collaboration, and small-scale entrepreneurship provides a valuable opportunity for a community to potentially receive economic benefits while simultaneously avoiding many of the problems associated with traditional approaches to economic development. Furthermore, these makerspaces provide a potential pathway to revitalize a city's downtown, mainstreet, or urban core without requiring significant infrastructure investments. These makerspaces can serve as a catalyst for additional development into declining downtowns and ultimately change the landscape of an entire area by encouraging people to return to declining areas. Furthermore, any business creation resulting from a makerspace will originate within the community and will likely be achieved without significant public investment into the typical financial incentives used to attract businesses for economic development purposes. As will be highlighted in the following case study, makerspaces are designed to be shared widely in a community and can also contribute to increases in social capital—a key dimension of social sustainability (Opp 2017). Finally, the maker movement aligns well with an innovation-focused economic development strategy that offers important synergies with the smart growth planning efforts of a city (NLC 2016).

Making Macon Sustainable: SparkMacon

Macon, Georgia, is a historically significant community located approximately 85 miles south of Atlanta. The city was founded in 1823 and is home to more than 6,000 historic buildings and 14 historic districts (Historic Macon 2017). The community's 155,000 citizens consist primarily of racial and ethnic minorities with approximately 43 percent of the population being classified as "white alone" in the 2015 American Community Survey. The community also has a relatively high poverty rate with almost a quarter of the city's families living below the poverty line. The median household income in 2015 was $36,568 and the unemployment rate was just slightly over

7 percent. Macon has struggled economically over the last several decades and by many accounts has not recovered from the Great Recession.

In 2012 after a tumultuous fight that had economic, government efficiency, and racial undertones, Macon City and Bibb County voters approved a city-county consolidation forming the new Macon-Bibb County Government (Stucka 2012). The consolidated government structure consists of an elected mayor that serves as the head executive and a 10-member commission serving as the legislative body for the community (Maconbibb. us 2017). According to local economic development commission data, the major employers in Macon include insurance, health care, and education. Other important industries also include manufacturing and distribution services (Gateway Macon 2013). The community is also home to five institutions of higher education with each playing some role in the local economy (Macon Colleges and Universities 2015). A 2014 study found that Macon had one of the slowest-growing economies in the nation and was continuing to struggle to return to prerecession employment and income levels (Stucka 2014). Furthermore, the 2004 departure of a large local tobacco company— Brown & Williamson—created a substantial economic hole in the community that has lingered as a significant economic challenge over the years (Rasmussen 2013). Brown & Williamson employed more than 3,000 people at its peak with an average pay of $26 per hour (Dewan 2004). The closure of this plant had ripple effects throughout the community as the company was one of the largest donors to local nonprofits and was also a significant contributor to local tax receipts (Dewan 2004). Between the closure of this major employer and the ongoing effects of the Great Recession, economic development and economic diversification has been an important focus in recent years and was, at least partially, responsible for the eventual approval of the city-county consolidation.

Although Macon is located within a state that is not particularly well known or well regarded for actively engaging in sustainability, this particular community has emerged as a potential unexpected leader on many important local sustainability efforts. Among the many examples of active local engagement into sustainability initiatives is the fact that the mayor of Macon joined the Mayors National Climate Action Agenda (MNCAA) affirming a commitment to combat climate change and reduce greenhouse gas emissions. Furthermore, in 2016, the local government expanded the local recycling program leading to a doubling of the community's recycling collection rates. Additionally, in recent years the local street lights have been upgraded to LED lights, an aggressive tree-planting effort has been adopted, electric vehicles have been added to the government's fleet, and parks and natural spaces have been expanded and improved upon across the community (Floore 2017). Finally, much of the economic development partnership work in Macon has included an explicit focus on encouraging vibrant, walkable, and diverse neighborhoods.

Macon's commitment to sustainability has carried over to the development and revitalization efforts of the downtown area. As the current mayor says, "Our efforts to become the Hub City of Middle Georgia—being a walkable city, attracting people back to Downtown, improving sidewalks and open spaces, encouraging mixed use development—are core tenants of revitalizing a community . . . and they can help lower our negative impact on the environment" (Floore 2017). In 2011, the local special purpose sales tax (SPLOST) provided an important starting point for downtown revitalization by contributing $8 million to begin the redevelopment of a portion of the downtown in order to make it pedestrian friendly and more sustainable (Rasmussen 2013). Ongoing and longer-term plans for downtown include a complete streets program, mass transit development, investment into an African American museum, enhancing public spaces, creating a bike-friendly downtown, and loft apartment living (Dewan 2004; Rasmussen 2013). Macon has also made extensive use of available technical assistance offered by Smart Growth America to further their interests in place-making, smart growth, and sustainable development (Smart Growth America 2017). As a very recent example of this entrepreneurial search for sustainable development assistance, in late 2017, the city participated in the "Amazing Place Ideas Forum" in order to glean insights and ideas about place-making and smart growth for their community. The community also maintains an important coalition of area nonprofits, such as the College Hill Alliance and the Knight Foundation, that are all focused on various aspects of sustainable development.

In light of the lingering economic problems associated with the Great Recession, Macon has aggressively focused their economic development efforts on place-based initiatives to try to "unlock the best version of Macon that we can be" (Smart Growth America 2017). The assistant manager for the Economic and Community Development Department sums up Macon's approach to economic development, "[Economic development] has come to mean jobs, but when we get down to what an economy actually is, it's how people interact . . . it's how do people interact, grow, and flourish together. That's what place-based strategies really are all about—how do the people experience a place" (Smart Growth America 2017). A focus on sustainability and quality of life is embedded in the public economic development efforts of this community. The maker movement has become a natural extension, and perhaps an important early accomplishment, of this emerging and expanding sustainability focus in Macon, Georgia.

The Emergence of the Makerspace: SparkMacon

In 1965, the counties of Bibb, Crawford, Houston, Jones, Monroe, Peach, and Twiggs created a voluntary planning commission called the "Middle Georgia Area Planning Commission" in order to provide development and planning technical assistance to the communities in the area (Middle Georgia Regional Commission 2017). Over the years this organization has expanded to include a larger and more significant role in the region's economic development efforts. The organization was ultimately designated as an Economic Development District (EDD) by the Department of Commerce in 1978 providing important access to federal economic development dollars. The most recent transition for this group came in 2009 when the organization was renamed the Middle Georgia Regional Commission (MGRC) (Middle Georgia Regional Commission 2017). MGRC's mission is to support comprehensive planning and economic development for the communities located within its jurisdiction—including Macon. It is this organization that led the research and the planning efforts that ultimately culminated in bringing the maker movement to middle Georgia (SparkMacon 2017).

The maker movement has been an important and visible area of focus for MGRC over the past several years. For example, in the 20-year comprehensive plan published in early 2016 makerspaces are explicitly mentioned as an important tool to "provide spaces for individuals to safely test ideas and strategies for new businesses ideas. If successful, some of these individuals may have the opportunities they need to help lift themselves out of poverty" (Middle Georgia Regional Commission 2016, 33). Furthermore, survey research conducted by MGRC has demonstrated the significant public support that makerspaces enjoy in the area with more than two-thirds of the survey respondents either agreeing or strongly agreeing that the region should encourage a "maker culture" as part of the local economic development effort. This ongoing focus on makerspaces eventually led to the creation of SparkMacon—the first makerspace in middle Georgia.

The actual process of launching SparkMacon was a collaborative cross-sectoral effort from the very beginning. Initially, five individuals representing MGRC, Mercer University, a local economic revitalization organization known as the College Hill Alliance, and two private sector representatives made up the founding board members. These original board members were responsible for turning the general idea of a makerspace in Macon into a reality for the community. This group was ultimately successful in opening SparkMacon in November 2014 thanks, in part, to some creative fundraising efforts. To obtain the seed money needed to open the space and purchase the necessary equipment, the five board members applied for, and received, a grant from the Georgia Technology Authority (GTA). Alongside the successful GTA grant application, the group also launched an impressive ground-level crowdfunding campaign that generated 111 percent of the fundraising goal (Rosario 2014). Alongside the fundraising campaign, the original board members were also aggressive in seeking out guidance, lessons, and ideas from other successful makerspaces in Georgia. These board members also spent time developing partnerships with other community members to help set the stage for a successful makerspace with significant community buy-in. Two months before officially opening, SparkMacon had the support and partnership of more than a dozen entities in the community ranging from nonprofits, to the chamber of commerce, to the local government (Inspired to Educate 2014). When the space finally opened in downtown Macon it was one of only 30 makerspaces in the entire state of Georgia and was the only one located in Central Georgia.

Present day, more than three years after opening, SparkMacon is a vibrant membership-driven makerspace with an eight-member board of directors responsible for the day-to-day management and long-term sustainability of the space (SparkMacon 2017). Rather than focus on the most common science, technology, engineering, and mathematics (STEM) areas, SparkMacon has intentionally developed what they call a "STEAM" focus with the "A" adding an emphasis on "Arts" that is often missing from makerspaces. The 3,000-square-foot makerspace has evolved to include nine distinct areas ranging from a wood and metal shop to a music studio. The space also includes a conference area, classroom space, and a lounge area for what they call "creative entrepreneurs" to share ideas and to potentially collaborate on projects. The current mission statement demonstrates the strong connection the space is meant to have to local economic development and entrepreneurship.

The mission of SparkMacon is to create a collaborative workspace for a wide range of creatives that meets every stage of entrepreneurship, from ideation to prototyping to manufacturing, by providing an adaptable space for innovation, in-house support, as well as connections to local resources that enable them to make a living doing what they love. (SparkMacon 2017)

Two key areas of sustainable development are closely aligned with makerspaces and are what makes these spaces a potential economic development tool for cities. First, entrepreneurship and the local production of goods provide a mechanism to focus local economic development efforts internally as opposed to more traditional needs-based external approaches to economic development. Many traditional approaches to economic development have been associated with negative outcomes for many cities—including environmental externalities (Koven and Lyons 2010; Opp and Osgood 2013). The internal focus associated with makerspaces can potentially avoid many of these negative outcomes. Second, human resource development and social capital can be naturally pursued through makerspaces and therefore contribute to local sustainability and economic goals. More will be said on each of these in the text that follows as it relates to SparkMacon.

Entrepreneurship and Local Production of Goods

Entrepreneurship and the local production of goods have been identified as a potentially important and significant sustainable development tool for cities (Dean and McMullen 2007; Pacheco, Dean, and Payne 2010; Shuman 2009). Although the encouragement of entrepreneurship has been a long-time approach to economic development, it has more recently expanded into the sustainable development literature through a focus on local production activities. Entrepreneurship based on local production has the potential to contribute to sustainability goals through "reduced transportation needs, lower pollution, more local jobs, local ownership of businesses, and retention of capital in the local community" (Wheeler and Beatley 2009, 233). By encouraging local production and entrepreneurship in downtown Macon, SparkMacon holds potential to shift the local economic development paradigm to one that is more innovative, competitive, and ultimately sustainable and resilient. SparkMacon provides the necessary shared space and educational support for all stages of entrepreneurship from the idea stage to the actual manufacturing process. Furthermore, SparkMacon has developed important ongoing partnerships in the community that can be used to help local makers bring a new product to market and eventually create a new business (SparkMacon 2017).

Human Resources Development and Social Capital

Alongside the entrepreneurship-based economic development potential, SparkMacon also contributes to human resource development through their ongoing educational and outreach activities. Several workshops and activities are held at SparkMacon each month with many of them being free to the public. For example, SparkMacon frequently hosts an "Introduction to Technical Training" class meant to teach basic skills in 3-D printing, laser cutting, coding, drones, and microcontroller programming. This type of educational opportunity represents an important tool for local human resource development and workforce training in the city. Furthermore, this type of workshop could provide the impetus for individuals within the community to choose to pursue additional education as a result of this exposure. "The nation is not preparing its human resources for competition today or tomorrow. . . . Even as jobs are being shed, people remain an important community resource and are crucial to a locality's economic recovery" (Leigh and Blakely 2016, 335). Introducing Maconites to these new and emerging technologies will, at minimum, provide an avenue for these residents to develop a skillset useful for the changing tech-based job market. Additionally, introducing at-risk youth to small-scaled manufacturing may also offer an avenue to encourage additional education and training pursuit within this community. Even better, however, might be the potential to spur innovation that ultimately leads to the birth of new local firms and a path for upward mobility and increased wealth for Macon residents. Important to also mention is that in addition to the training workshops offered to the general public, SparkMacon has added a distinct social dimension to some of their outreach and training programs. For example, programming for displaced workers and veterans is emphasized in some of their technical training courses. Additionally, SparkMacon has also engaged in a partnership with an all-girl STEM focused group, Real Impact, to support the encouragement of females in STEM and entrepreneurship (Evans 2015). More recently, in mid-2017, SparkMacon received a Department of Defense grant to help provide an avenue to develop skills for displaced veterans (Nadia Osman, personal communication).

Alongside the important contributions to human resource development, SparkMacon's focus on workshops and activities also hold some promise for another important sustainability dimension known as social sustainability. One part of pursuing social sustainability involves a focus on the social capital stock of a community (Opp 2017). Research has provided evidence of the importance and economic value of actively encouraging and facilitating an increase in social capital stock within a community (Knack and Keefer 1997; Opp 2017; Putnam 1995; Selman 2001. Although social capital is defined many ways, most definitions of this concept refer to the networks, trust, and relationships that exist (or don't exist) within a community (Putnam 1995). Higher levels of social capital are thought to exist when trust, collaboration, and cooperation is clearly present and encouraged within a community. A high level of social capital stock is associated with healthier economies and more successful sustainability efforts (Rupasingha, Goetz, and Freshwater 2000; Opp 2017). SparkMacon is providing an important venue for the development of the building blocks of social capital in Macon, Georgia. Through the shared makerspace, people who otherwise would not talk or work with each other are able to come together in a creative and safe space to collaborate and build relationships through their maker experiences.

ECONOMIC IMPACT

Although the full economic impact of SparkMacon has not been formally assessed, several important related activities have emerged that provide some useful evidence of the early economic successes this space is having. As one example, in July 2016, the Downtown Macon Open Air Market was formed and has now grown to include over 60 vendors. This market is currently held on a quarterly basis in downtown Macon and has provided a public venue for local makers, including those associated with SparkMacon, to sell their products to area residents (Gateway Macon 2017). As another example of the economic impact SparkMacon is having on Macon, a local cosplay group was able to use the shared space to use the laser cutters and woodworking shop to create needed costumes and props for their successful video project (Evans 2015). The inclusion of the "A" in "STEAM" is at least partially responsible for the creation of two other makerspace-like facilities in the community: & Guild that focuses on art and artists and the music incubator known as 5/4 Music Space (Nadia Osman, personal communication). Finally, perhaps as a more direct and traditional connection to economic development goals, SparkMacon has contributed to the launch of several new local businesses in Macon—some of which are now operating a storefront and some of which are still housed by SparkMacon.

Among the various businesses launched with some help from SparkMacon is the minority-owned custom bow tie business BowFRESH and the unique furniture reclamation business Georgia Artisan (Middle Georgia

CEO 2017). As the owner of BowFRESH says about SparkMacon, "The space has provided a comfortable atmosphere to make my handcrafted bowties, as well as given me access to amazing technology resources, including a laser cutter and photo editing software" (Middle Georgia CEO 2017). BowFRESH is a prime example of the positive impact that a makerspace can have on both the community and on the local residents. Aaron Brown, the owner of BowFRESH, was in his early 20s and working in a major department store when the store closed leaving him unemployed and searching for creative ways to find an income stream in an economically struggling community (Bennett 2016). SparkMACON was instrumental in providing him with a space, the equipment, and the support needed to take his creativity and interest in sewing to the level needed to launch a business of his own. By many accounts, Georgia Artisan is likely the easiest success story to point to when looking at the impact SparkMacon has made. The founder, Andrew Eck, now operates his own warehouse employing five people and got his start in business through SparkMacon (Nadia Osman, personal communication).

CONCLUSIONS AND MOVING FORWARD

Macon, Georgia, is not a community that most people immediately think of when they are searching for examples of successes in local sustainability efforts. However, the substantial local investment into learning, growing, and being creative with sustainable development techniques is beginning to pay dividends for the community. Rather than engage in some of the traditional needs-based externally focused approaches to economic development, Macon has intentionally and methodically focused efforts internally and on ways to balance sustainability with their very real economic needs. This impressive engagement in sustainability initiatives in a state that generally does not provide the kind of support that many other states do is remarkable and instructive to learn from. Furthermore, Macon, Georgia, does not enjoy many of the geographic perks that the typical leaders of sustainability across the country have. For example, cities like San Francisco, Portland, and Boulder are often hailed as exemplars in the local sustainability movement. However, unlike these communities, Macon does not enjoy the tourism (and therefore economic benefit) draw that the Rocky Mountains or the San Francisco Bay provides for these oft-labeled sustainable cities. Furthermore, Georgia does not provide the same level of incentives to their cities to engage in sustainability policymaking as do states like California or Colorado. Macon is located in a poor state, with limited state support, lagging rates of higher education, and significant economic constraints. However, with some planning, creativity, persistence, and dedication to the principles of sustainability, they have made some very real progress on sustainably revitalizing their community.

Makerspaces are relatively new to the conversations on sustainability, and much remains to be studied in order to fully understand the role they play in this policy area. However, these bottom-up collaborative focused makerspaces show great promise for contributing to local sustainable economic development in American cities. In many communities across the United States downtown revitalization remains both a priority and a challenge. As shopping malls and retail establishments continue to decline across the country, communities must find new and creative ways to reuse these now vacant or underutilized spaces. Furthermore, cities must also find creative ways to provide for the very real economic needs of their residents. Makerspaces offer a relatively simple and inexpensive method of reusing underutilized downtown space while also providing for social capital and economic development needs. SparkMacon was able to open with relatively little in public financial support—primarily in the form of a state grant—and has already shown real economic impact through the creation of several new businesses. Modest membership fees cover the basic operating expenses, and the community does not have to invest significant dollars into making a makerspace work. Macon, Georgia, will be a valuable and important case to watch in the years ahead as they continue their aggressive sustainability push under some of the least supporting conditions in America. This is truly an example of "if this city can do it then any city can do it."

REFERENCES CITED

Bennett, Josephine. 2016. "Macon Entrepreneur Making Fashion." GPB Media. July 26. Last accessed November 3, 2017. http://www.gpb.org/blogs/right-here/2016/07/26/macon-entrepreneur-making-fashion.

Brundtland. 1987. Report of the World Commission on Environment and Development: Our Common Future. United Nations.

Dean, Thomas J., and Jeffery S. McMullen. 2007. "Toward a Theory of Sustainable Entrepreneurship: Reducing Environmental Degradation Through Entrepreneurial Action." Journal of Business Venturing 22(1): 50–76.

Deslatte, Aaron, and Eric Stokan. 2017. "Hierarchies of Need in Sustainable Development: A Resource Dependence Approach for Local Governance." Urban Affairs Review. DOI: 1078087417737181.

Dewan, Shaila. 2004. "Its Main Employer Leaving, Macon Tries to Diversify." New York

Times, July 17. http://www.nytimes.com/2004/07/17/us/its-main-employer-leaving-macon-tries-to-diversify.html.

EPA. 2017. "Local Foods, Local Places." Last modified October 2, 2019. https://www.epa.gov/smartgrowth/local-foods-local-places.

Evans, Megan. 2015. "Spark Macon: A Maker's Space in Downtown Macon." 11th Hour, June 29.

Floore, Chris. 2017. "Macon-Bibb Joins the Mayors National Climate Action Agenda." June 16. http://www.maconbibb.us/climatemayors/.

Gateway Macon. 2013. "Major Employers." Last accessed January 28, 2020. https://www.gatewaymacon.org/about-macon-ga/major-employers.cms.

Gateway Macon. 2017. "Downtown Macon Open Air Market." Last accessed November 10, 2017. https://www.gatewaymacon.org/things-to-do-macon-ga/gatekeepers-top-picks.cms?id=2487.

Historic Macon. 2017. "Macon's Historic Districts." Last accessed November 2, 2017. http://www.historicmacon.org/macons-historic-districts/.

Inspired to Educate. 2014. "SparkMacon: Our MakerSpace for Macon." Last accessed November 6, 2017. http://inspiredtoeducate.net/inspiredtoeducate/sparkmacon-our-makerspace-for-macon-ga/.

Kalil, Thomas. 2017. "Building a Nation of Makers." Last modified September 13, 2017. https://www.huffpost.com/entry/building-a-nation-of-make_b_11978108.

Knack, Stephen, and Philip Keefer. 1997. "Does Social Capital Have an Economic Payoff? A Cross-Country Investigation." Quarterly Journal of Economics 112(4): 1251–1288.

Koven, Steven, and Thomas Lyons. 2010. Economic Development: Strategies for State and Local Practice. Washington, D.C.: International City/County Management Association.

Leigh, Nancey Green, and Edward Blakely. 2016. Planning Local Economic Development: Theory and Practice. Sage Publications.

Maconbibb.us. 2017. Last accessed October 21, 2017. http://www.maconbibb.us/.

Macon Colleges and Universities. 2015. Last accessed November 2, 2017. https://www.maconga.org/about/colleges-universities/.

Middle Georgia CEO. 2017. "SparkMacon Launches Collaborative Workspace." Last accessed October 10, 2017. http://middlegeorgiaceo.com/news/2017/10/sparkmacon-launches-collaborative-workspace/.

Middle Georgia Regional Commission. 2016. 2016–2036: Plan for . . . a Thriving Middle Georgia. https://www.middlegeorgiarc.org/wp-content/uploads/2015/09/Regional-Agenda_Draft020416.pdf.

Middle Georgia Regional Commission. 2017. Last accessed November 2, 2017. https://www.middlegeorgiarc.org/about-us/.

NLC. 2016. "How Cities Can Grow the Maker Movement." Last accessed October 22, 2017. http://www.nlc.org/sites/default/files/2016-12/Maker%20Movement%20Report%20final.pdf.

Opp, Susan. 2017. "The Forgotten Pillar: A Definition for the Measurement of Social Sustainability in American Cities." Local Environment 22(3): 286–305.

Opp, Susan, and Jeffery Osgood Jr. 2013. Local Economic Development and the Environment: Finding Common Ground. Boca Raton: CRC Press.

Opp, Susan, Jeffery Osgood Jr., and Cynthia Rugeley. 2014. "City Limits in a Postrecessionary World: Explaining the Pursuit of Developmental Policies After the Great Recession." State and Local Government Review 46(4): 236–248.

Osgood, Jeffery, Susan Opp, and Megan DeMasters. 2017. "Exploring the Intersection of Local Economic Development and Environmental Policy." Journal of Urban Affairs 39(2): 260–276.

Osgood Jr., Jeffery, Susan Opp, and Laurie Bernotsky. 2012. "Yesterday's Gains Versus Today's Realties: Lessons from 10 Years of Economic Development Practice." Economic Development Quarterly 26(4): 334–350.

Pacheco, Desiree F., Thomas J. Dean, and David S. Payne. 2010. "Escaping the Green Prison: Entrepreneurship and the Creation of Opportunities for Sustainable Development." Journal of Business Venturing 25(5): 464–480.

Putnam, Robert D. 1995. "Bowling Alone: America's Declining Social Capital." Journal of Democracy 6(1): 65–78.

Rasmussen, Patty. 2013. "Macon Bibb County: Consolidation and Diversity." Georgia Trend, March. http://www.georgiatrend.com/2013/03/01/Macon-Bibb-County-Consolidation-And-Diversity/.

Rosario, Michael. 2014. "Our Crowdfunding Campaign Has Launched." Last accessed November 1, 2017. https://www.sparkmacon.com/our-crowdfunding-campaign-has-launched-contribute-now-and-tell-your-friends/.

Rupasingha, Anil, Stephan Goetz, and David Freshwater. 2000. "Social Capital and Economic Growth: A County-Level Analysis." Journal of Agricultural and Applied Economics 32(3): 565–572.

Selman, Paul. 2001. "Social Capital, Sustainability and Environmental Planning." Planning Theory & Practice 2(1): 13–30.

Shuman, Michael. 2009. "Import Replacement." In Sustainable Urban Development Reader, edited by Stephen Wheeler and Timothy Beatley. London: Routledge.

Smart Growth America. 2017. "Amazing Places Series: 'A Change Is Gonna Come' in Macon, Georgia." Last accessed November, 2017. https://smartgrowthamerica.org/amazing-place-series-change-gonna-come-macon-georgia/.

SparkMacon. 2017. "SparkMacon." Last accessed January 27, 2020. https://www.sparkmacon.com/.

Stucka, Mike. 2012. "Macon-Bibb County Consolidation Wins with Strong Majorities." Telegraph, July 31. http://www.macon.com/news/politics-government/election/article30109740.html.

Stucka, Mike. 2014. "Report: Macon's Economy to Grow Among Slowest in U.S." Telegraph, June 23. http://www.macon.com/news/local/article30133248.html.

Wheeler, Stephen, and Timothy Beatley, eds. 2009. Sustainable Urban Development Reader. London: Routledge.

III

Focus on Atmospheric Sustainability in the Suburbs

9

Ozone in Urban North Carolina

A Sustainability Case Study

WILLIAM H. BATTYE, CASEY D. BRAY,
PORNPAN UTTAMANG, AND VINEY P. ANEJA

Ground-level ozone (or smog) is a common air pollution problem in urban areas, caused by air pollution emissions from automobiles, power generation, and other sources. In the 1970s, population growth and economic growth in North Carolina resulted in increased air pollution emissions and elevated levels of smog. Over the decades since then, economic growth has continued in North Carolina. However, air pollution controls have enabled the state to sustain this growth while reducing levels of urban smog. This chapter describes how the control of ozone has been achieved in the state.

Ground-level ozone is a secondary air pollutant that forms through the chemical reactions of its precursors in the atmosphere in the presence of sunlight (Aneja et al., 1992, 39–44). Exposure to ground-level ozone has both short-term and long-term adverse effects on human health, especially effects on respiratory system. The severity of ground-level ozone on human health depends on its concentration and the physical health of receptors. Children, elders, people with health problems, and active people are more susceptible to disease and illness from ground-level ozone than other groups. Adverse health impacts from ground-level ozone exposure can be eye irritation, respiratory system and lung irritation, reduced lung function and cell damage, asthma, cardiovascular problems, and mortality.

Ground-level ozone can damage plants and material. Plants exposed to ozone express several symptoms such as chlorosis and necrosis, reducing or stop growth, and reducing crop yields. Exposing ozone can be a cause of rubber cracking and deterioration of tires ("Effects of Ozone Air Pollution on Plants" 2016).

As opposed to ground-level ozone, stratospheric ozone protects human health and life on the earth by absorbing ultraviolet (UV) radiation. UV radiation is electromagnetic wave with short wavelength (λ) ranges from 100 to 400 nm. UV is separated into three types based on their wavelengths— UVA ($\lambda \sim$ 320–400 nm.), UVB ($\lambda \sim$ 290–320 nm.) and UVC ($\lambda \sim$ 100–290 nm.). The shorter the wavelength radiation, the higher the energy it carries. Only UVA and a small amount of UVB can penetrate into the lower troposphere; other, shorter wavelengths are absorbed by ozone in the atmosphere, especially in the stratosphere. Because the stratospheric ozone absorbs UV radiation that can cause eye and skin irritation, reduce immune function, and cause skin cancer, it has been known as "good" ozone ("Ozone and Your Health" 2009). On the other hand, the ground-level ozone is harmful to both human health and the environment and is therefore known as "bad" ozone.

History of Ground-Level Ozone Standard and Air Quality Index

To protect human health and environment from the adverse effects of ozone, the U.S. Environmental Protection Agency (EPA) has included ground-level ozone among the six criteria air pollutants regulated by National Ambient Air Quality Standards (NAAQS). The ground-level ozone standard has been set and revised on the basis of scientific information since 1971. In 1971, the chemical designation for NAAQS was total photochemical oxidant, and it was changed to hourly ozone concentration in 1979. The ground-level ozone standard in 1979 was 0.12 part per million (ppm) for one hour. However, the EPA undertakes periodic reviews of NAAQS, and these reviews have resulted in a gradual reduction of the level of the standard. In 2014, the ground-level ozone standard was set at 0.07 ppm, averaged over eight hours, an almost 50-percent reduction compared with the standard in 1979 ("Ozone Trends" 2016).

Ground-Level Ozone Chemistry

In urban areas with a high concentration of pollutants, photochemical reaction of nitrogen oxide (NO_x, $NO_x = NO + NO_2$) with peroxy radicals (RO_2), is the major pathway to produce ground-level ozone in the atmosphere. The chemical reactions ((R1)—(R3)) illustrated in Figure 9.1 show the ground-level ozone formation in urban area.

$$NO\ (g) + RO_2\ (g) \rightarrow NO_2\ (g) + RO\ (g)\quad (R1)$$

$$NO\ (g) + HO_2\ (g) \rightarrow NO_2\ (g) + OH\ (g)$$

$$NO_2\ (g) + h\upsilon\ (\lambda \leq 420\ nm) \rightarrow NO\ (g) + O\ (g)\quad (R2)$$

$$O\ (g) + O_2\ (g) + M \rightarrow O_3\ (g) + M\quad (R3)$$

Figure 9.1. Chemical reactions for ground-level ozone formation in urban areas.

The major sources of NO_x emission are motor vehicles, industrial processes, and electric power plants. Volatile organic compounds (VOCs) and reactive organic gases (ROGs) are emitted from both natural sources as well as human activities. In urban areas, the major sources of VOCs and ROGs emission are anthropogenic sources, that is, solvent usage, fuel combustion, industrial processes, on-road and off-road vehicles and engines; and natural sources, that is, trees (Jacobson 2012, 73).

Certain meteorological conditions may enhance ground-level ozone accumulation. For example, high-pressure systems and low wind speeds enhance ground-level ozone accumulation. In addition, high temperature, high solar radiation, and low humidity increase ground-level ozone formation and increase its concentration in the atmosphere ("Trends in Ozone Adjusted for Weather Conditions" 2016).

EFFECTS OF URBANIZATION ON GROUND-LEVEL OZONE CONCENTRATION

Civerolo et al. (2007) studied the effects of increased urbanization in New York City (NYC) coupled with meteorology and ground-level precursor concentration by simulation models. The study showed the increased urbanization had the potential to increase temperature and mixing height but decrease water mixing ratio in the atmosphere. Increased urbanization has a greater impact on ground-level ozone concentration than meteorology. Wang et al. (2007) showed increasing industrial and urban land-use types increased temperature and mixing height in the Perl River Delta, China, but decreased wind speed in the region.

In this chapter, we present the basic knowledge of ground-level ozone concentration including causes and effects, regulation, and mitigations to control ground-level ozone in the atmosphere by focusing on the Triangle, Triad, and Charlotte regions of North Carolina.

Economic Growth and Air Pollution Emission Trends in North Carolina

Over the last half century, North Carolina has been one of the fastest-growing states in the U.S. The state's population has doubled since 1970 ("Current Lists of Metropolitan and Micropolitan Statistical Areas and Delineations" 2016). Much of this growth has occurred in the cities and suburbs of the Piedmont region of central North Carolina. This region extends from the Charlotte metropolitan area in the west to the Raleigh metropolitan area in the east. Both of these cities are among the ten fastest-growing Metropolitan Statistical Areas (MSA) in the U.S. Raleigh, Durham, and Chapel Hill make up North Carolina's Research Triangle, a center of technological, medical, and pharmaceutical research. Charlotte is the 17th-largest city in the U.S., and a major financial and corporate center ("Census of Population and Housing" 2016; "Annual Estimates of the Resident Population 2015 Population Estimates" 2016). Between the Research Triangle and Charlotte is the Triad region. This region encompasses the cities of Greensboro, High Point, and Winston-Salem, with a total metropolitan population of about 1.6 million. The Triad is a known center of textile and furniture manufacturing, and although these industries have declined, the region remains a national center. In addition, the textile and furniture industries drove the development of a trucking and warehousing infrastructure which remains active.

The population of North Carolina has more than doubled since 1960, and economic activity in the state has increased faster than the population. Energy consumption can be used as an indicator of economic activity. Figure 9.2 shows the growth in population, electricity consumption, and vehicle miles traveled (VMT) since 1960. It also shows the growth in industrial and commercial/institutional energy use. Electricity usage and VMT have increased almost six-fold since 1960. Demand for electricity rose from 203 million kilowatt-hours in 1960 to 1,176 million kilowatt-hours in 2014. VMT in the state has increased from about 19 million to about 108 million (Figure 9.2) ("Highway Statistics, 1960–2014" 2016; "Census of Population and Housing" 2016; "Annual Estimates of the Resident Population 2015 Population Estimates" 2016). Industrial energy demand has remained somewhat stable, while energy demands in the consumer and commercial sectors have increased at a faster pace than population (Figure 9.2) ("State Energy Data System (SEDS): 2016–2014—North Carolina" 2016; "Census of Population and Housing" 2016; "Annual Estimates of the Resident Population 2015 Population Estimates" 2016).

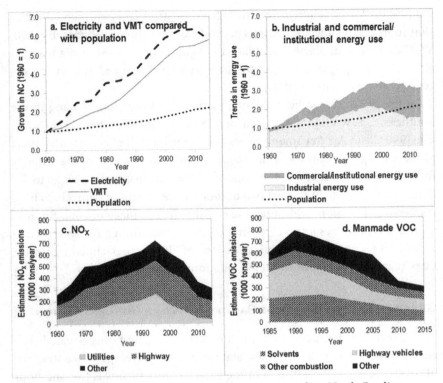

Figure 9.2. Economic growth and air pollution emission trends in North Carolina.

In the absence of control measures, man-made air pollution emissions would generally follow the underlying trends in population and economic growth, and this was initially the case with NO_x and VOC in North Carolina. Figure 9.2 shows the increases in emissions of these pollutants in the state, broken down by major emissions source types. NO_x emissions are estimated to have almost tripled between 1960 and 1996, from 257,000 tons per year to 739,000 tons per year ("Air Pollutant Emissions Trends Data, State Average Annual Emissions Trend, 1990–2014" 2016; Gschwandtner et al. 1985, 139–149). Data are not available on VOC emissions from 1960 to 1985, but emissions are estimated to have increased by about 30 percent between 1985 and 1990, from 607,000 to 796,000 tons per year ("Air Pollutant Emissions Trends Data, State Average Annual Emissions Trend, 1990–2014" 2016).

The Clean Air Act amendments of 1970, which established the U.S. EPA, also mandated the development of federal and state regulations for the control of air pollution, including urban smog. At first, these regulations focused on VOC, requiring controls for industrial and commercial emission sources, and lower emissions per mile for new motor vehicles. By 1990, VOC emissions in North Carolina had begun to decline. Man-made emissions of anthropogenic VOC in the state have been reduced by over 60 percent, to 304,000 tons in 2014.

In the 1980s, researchers recognized that the mitigation of smog in the Southeast U.S. would require control of NO_x (Chameides 1988, 1473–1475). The 1990 Clean Air Act amendments mandated control of NO_x, both as part of the strategy for control of urban smog and also as a part of the strategy for curbing acid rain; thus, emissions of NO_x in North Carolina began to decline in about 1996. In the last two decades, lower prices of natural gas have also resulted in more widespread use of this fuel in place of coal. Because natural gas produces less NO_x than coal, this substitution has also served to reduce NO_x emissions. Overall, emissions of NO_x have declined by almost 60 percent since 1996, to about 307,000 tons per year in 2014.

General Trends in North Carolina Ozone Concentration

The state of North Carolina was plagued with smog, and thus elevated concentrations of ozone, for decades. Statewide hourly average concentrations of ozone were around 120 ppb in the early 1970s, which was the one-hour average ozone NAAQS set by the U.S. EPA from the early 1970s until 1997. By the late 1980s, the average one-hour concentration of ozone across North Carolina had reduced to 100 ppb. Statewide average one-hour concentrations of ozone remained fairly steady near 100 ppb while the newly measured eight-hour average concentrations of ozone hovered between 80 and 90 ppb throughout much of the 1990s. By the mid-1990s, much of North Carolina had reached attainment (i.e., the concentration of ozone is within the limits set by the U.S. EPA NAAQS) or maintenance (i.e., an area that exceeded the standard then has reduced and maintained one-hour average ozone concentrations such that they no longer exceed the standard) status for ozone for the one-hour average (*1972–1995 Ambient Air Quality Trends Summary* 1998, 7–19).

In 1997, the U.S. EPA discontinued the one-hour average ozone NAAQS in favor of an eight-hour average standard. This 1997 eight-hour average NAAQS for ozone was set to 85 ppb. Unlike the previous standard, North Carolina had a much more challenging time obtaining attainment, with the statewide average eight-hour concentration of ozone rising well above

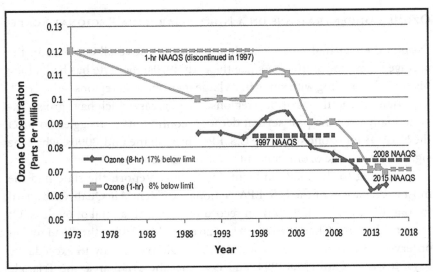

Figure 9.3. The statewide average ozone concentration trends from 1973 to 2015. (Source: NC DEQ Air Quality Trends in North Carolina Report, 2016.)

the 1997 standard, with average eight-hour concentrations peaking near 95 ppb in the early 2000s. This trend in the average ozone concentration can be seen in Figure 9.3, which shows the trend in average ozone concentrations (both the one-hour average and the eight-hour average) from the early 1970s to the present time and compares this trend with the 1979 one-hour ozone NAAQS and the 1997 eight-hour average ozone NAAQS as well as the 2008 eight-hour average ozone NAAQS set by the U.S. EPA. Despite the negative trend in ozone concentration and thus smog, it was still afflicting much of the state. Fortunately, the peak in poor air quality resulting from tropospheric ozone that occurred roughly around 2002/2003 marks the time when North Carolina's air quality started to change for the better. The U.S. EPA then reduced the eight-hour average concentration of ozone in 2008 to 75 ppb and then again in 2015 to 70 ppb, thus forcing some parts of the state back into nonattainment. However, it is important to note that while the state of North Carolina has continued to struggle to obtain clean, healthy air, the overall concentration of ozone, and thus smog, did decrease through the late 2000s up until the present day (*2001 Ambient Air Quality Report* 2008, 35–65; *2005 Ambient Air Quality Report* 2010, 31–83; *2011 Ambient Air Quality Report* 2013, 28–81).

Ozone Concentrations on a North Carolina Regional Scale

Despite the fact that North Carolina has seen an overall decrease in the average ozone concentration since the early 1970s, the state has had a long history of struggling with smog in its large metropolitan regions—the Triad, the Triangle, and the Charlotte regions. These areas, which have grown exponentially over the past several decades, continuously struggled to meet the NAAQS for ozone set by the U.S. EPA through the mid-2000s, when the eight-hour average ozone concentrations peaked.

According to the 1988 Ambient Air Quality Report, the state had the most exceedances of the U.S. EPA National Ambient Air Quality Standard for one-hour ozone (120 ppb) on record in 1988 with a total of 69 days. Of these 69 days, the Charlotte region saw 26 exceedance days, the Triad region observed 23 exceedance days, and the Triangle region saw 16 exceedance days. It wasn't until 1993, 1994, and 1995 that the Triad area, the Triangle area, and the Charlotte area, respectively, were redesignated to attainment from nonattainment ("Green Book 1-Hour Ozone (1979) Area Information—NAAQS Revoked" 2018). However, exceedances of this standard still occurred. According to the North Carolina Division of Air Quality's Annual Air Quality report for 1997, there were a total of four exceedances of the one-hour ozone standard that year. These exceedances occurred in the Triangle region, the Triad region, and the Washington region of North Carolina ("North Carolina Counties with 8-Hour Ozone Violations" 1997–1999).

In 1997, the U.S. EPA introduced a new eight-hour average standard (80 ppb), which stated that a region was in exceedance of the standard if the three-year average of the annual fourth-highest value of eight-hour average ozone concentration exceeded 80 ppb. This quantity (the three-year average of the fourth-highest eight-hour average) is also termed the ozone design value. Framing the standard in these terms reduces the impact of fluctuations caused by unusual meteorological conditions and thus more accurately reflects actual recurring pollution levels.

The new eight-hour standard put both the Triangle and the Charlotte region into nonattainment, with much of the state continuing to observe exceedances of the eight-hour average ozone NAAQS. From 1997 to 1999, there were 22 counties (with at least one county in each region) with ozone design values exceeding the eight-hour standard. Much of the state continued to see elevated eight-hour average ozone concentrations through the end of the 1990s and into the early 2000s. Between 2000 and 2002, there were 23 counties exceeding the NAAQS for eight-hour average ozone concentrations, with each region in violation. According to the NC DAQ Annual Air Quality Report for 2002, the eight-hour standard was exceeded 613 times during the 51-day ozone season. Fortunately for North Carolina, this was a turning point for air quality, with average eight-hour ozone concentrations beginning to decrease. By 2005, there were only three counties (within the Charlotte and Triangle regions) in exceedance of the NAAQS. In comparison to the 613 exceedances that occurred in 2002, there were only 75 exceedances of the eight-hour average ozone standard in 2005. By 2007, not only had the Triangle region been redesignated to maintenance/attainment, but there were only two counties (in the Charlotte region) that exceeded the NAAQS for eight-hour ozone for 2005–2007 (*Redesignation Demonstration and Maintenance Plan for the Charlotte-Gastonia-Salisbury, North Carolina 2008 8-Hour Ozone Marginal Nonattainment Area* 2015). However, the number of individual exceedances did marginally increase from 75 in 2005 to 79 in 2007.

As mentioned above, the U.S. EPA tightened the standards for average eight-hour ozone concentrations even further in 2008, where the standard is exceeded when the design value is greater than 75 ppb. Despite lowering the standard, only the Triad and Charlotte regions exceeded the new three-year average maximum eight-hour ozone concentration for 2008–2010. The eight-hour standard of 75 ppb was exceeded a total of 108 times during the 2010 ozone season, with the Charlotte region continuing to observe the highest ozone concentrations. However, smog was still an issue of concern in 2010 in the Charlotte region, with the area ranking 17th among all major U.S. metropolitan areas studied for the number of unhealthy days. In addition to this, the Charlotte region was ranked number 10 (out of 277) for the worst ozone days (*Charlotte Business Journal*, September 21, 2011).

By 2011, only the Charlotte region remained in nonattainment, with two counties exceeding the average eight-hour ozone standard for 2009–2011. However, it is important to note that the eight-hour standard of 75 ppb was exceeded a total of 100 times during the 2011 ozone season (*2011 Ambient Air Quality Report* 2013, 28–32). Then, in 2013, the air quality in North Carolina improved considerably, with only the Charlotte region exceeding the average eight-hour ozone concentration standard for 2011–2013. In fact, there was only one eight-hour average ozone concentration that exceeded 75 ppb in 2013 and that monitor was located in the Triad region. Since 2013, there have been a few (>10 per year) monitors with eight-hour ozone concentrations exceeding 75 ppb. However, the air quality has cleaned considerably compared to concentrations observed over the last several decades. While the air is still not entirely clean, the smog situation has improved significantly since 2002.

REGULATIONS ON AIR POLLUTION

Ozone concentrations have decreased over the past few decades (Figure 9.3). While NO_x and VOCs serve as precursor gases for ozone in the troposphere, North Carolina is a NO_x-limited regime (i.e., there is an abundance of VOCs thus NO_x controls are most effective for air pollution reduction). Therefore, reductions in NO_x emissions have really helped improve air quality in North Carolina. These reductions in emissions are primarily due to regulations set on air pollution at the local, state and federal levels.

Federal Level Regulations

Following the Clean Air Act, which was established in 1970 and had major revisions made in 1977 and in 1990, the U.S. EPA set several requirements to help improve air quality. These requirements include setting the National Ambient Air Quality Standards for the six criteria pollutants (particulate matter, sulfur dioxide, nitrogen dioxide, carbon monoxide, lead, and ozone) as well as requiring states to create plans in order to achieve and maintain these standards. In addition to this, there have been several federal actions that have been implemented throughout the eastern United States that have resulted in lower emissions of pollutants. Federal actions placed upon on-road and non-road vehicles include the Tier 2 engine standards for both light- and medium-duty vehicles, heavy-duty engine standards, and low-sulfur gasoline and diesel requirements as well as off-road engine standards. These standards placed on vehicles aim to lower the amount of pollutants, including NO_x, that these vehicles emit into the atmosphere. Moreover, federal actions that have been taken on stationary sources include the mercury and air toxics (MATS) rule that has been established for electricity generating units (EGUs) and the National Emissions Standards for Hazardous Air Pollutants (NESHAP) for industrial, commercial and institutional boilers and reciprocating internal combustion engines (RICE). Similar to the purpose of the on and off-road vehicle standards, both the MATS rule and the NESHAP also aim to reduce overall emissions of several pollutants, including NO_x. The most recent implementations made by the federal government include the Cross-State Air Pollution Rule (CSAPR) and the Tennessee Valley Authority consent decree. The purpose of the Cross-State Air Pollution Rule, which started in May 2017, is to reduce the NO_x emissions in the summertime (May–September) from 22 states in the eastern United States. While North Carolina is not included in these 22 states, this rule will reduce the formation of ozone in other areas and will therefore reduce the ozone pollution that crosses state lines into North Carolina ("Final Cross-State Air Pollution Rule Update" 2016; "Fact Sheet: Final Cross-State Air Pollution Rule Update for the 2008 NAAQS" 2016).

STATE-LEVEL REGULATIONS

North Carolina has implemented several regulations that have led to cleaner, healthier air across the state. These regulations include the NC Air Toxics Control Program, the heavy-duty diesel engine gap filling requirements, NO_x State Implementation Plan (SIP) Call rule, the Clean Smokestacks Act, the onboard diagnostic (OBDII) vehicle inspection and maintenance program as well as a voluntary program to reduce emissions from diesel engines (*Section 110(a) Maintenance Plan for the 1997 8-Hour Ozone Standard for Greensboro/Winston-Salem/High Point 1-Hour Ozone Maintenance Area* 2013; *Redesignation Demonstration and Maintenance Plan for the Charlotte-Gastonia-Salisbury, North Carolina 2008 8-Hour Ozone Marginal Nonattainment Area* 2015; "Key Facts About the Clean Smokestacks Act" 2016). The purpose of the North Carolina Air Toxics Control Program is to reduce toxic air pollutants, including volatile organic compounds. Established in 1990, this required all new facilities and existing facilities that are modified to comply with the air toxic rules before beginning operation. The heavy-duty diesel engine gap filling requirements of North Carolina was a rule adopted in 2001 by the Environmental Management Commission in order to fill a gap in the federal requirements for improved heavy-duty diesel engine emissions testing. This was established in order to prevent excess NO_x emissions from occurring over the lengthy life of these vehicles. This rule requires all 2005- and 2006-model-year heavy-duty diesel vehicles within North Carolina must be the type certified by the California Air Resources Board. The NO_x SIP Call rule was set to reduce NO_x emissions from North Carolina's large stationary combustion sources by 68 percent by 2006. The OBDII vehicle inspection and maintenance program requires emission inspections in 48 of North Carolinas counties and the on-board diagnostics system alerts the driver as well as the technician to a problem before any significant emission increase occurs. However, among these regulations implemented, the Clean Smokestack Act has been the most effective. The Clean Smokestack Act, which was enacted into legislation in 2002, required North

Carolina power plants to reduce emissions of NO_x and SO_2 by 77 percent by 2009 and 73 percent by 2013, respectively. ("Key Facts About the Clean Smokestacks Act" 2016). It is important to note that this act is different from the federal rules because it requires an actual reduction in emissions as opposed to allowing the utilities to buy pollution credits from other states. This act was a key contributor in reducing emissions of NO_x, among other pollutants, into the atmosphere and thus reducing ozone levels across the state.

Local-Level Regulations in Urban North Carolina

Due to its rapidly growing population, the Charlotte region has struggled the most with air quality improvements in North Carolina. However, this region has done a lot to try to improve its pollution levels. On a local level, the Charlotte region (controlled by Mecklenburg County Air Quality) also implemented several regulations in order to better the region's air quality. These implementations include a prohibition on open burning and a voluntary program called Grants to Replace Aging Diesel Engines (GRADE), which is designed to reduce the emissions of NO_x into the atmosphere by providing incentives to local businesses for replacing heavy-duty non-road equipment with more environmentally friendly engines (*Redesignation Demonstration and Maintenance Plan for the Charlotte-Gastonia-Salisbury, North Carolina 2008 8-Hour Ozone Marginal Nonattainment Area* 2015).

In the Triangle region, a ban was placed on open burning on ozone action days (days when ozone concentrations have the potential to elevate into the Code Orange range or above). In addition to this, the combination of state and federal regulations, such as the Clean Smokestack Act and the regulations placed upon vehicles, have helped significantly improve air quality in the region (*2011 Ambient Air Quality Report* 2013, 81).

There were no major local-level regulations that were implemented within the Triad area as the combination of the federal and state regulations were sufficient to improve air quality. In particular, the Clean Smokestack Act had a major impact in this area due to the presence of a large coal-fired power plant within the region (*Section 110(a) Maintenance Plan for the 1997 8-Hour Ozone Standard for Greensboro/Winston-Salem/High Point 1-Hour Ozone Maintenance Area* 2013).

SUSTAINABILITY AND MAINTENANCE OF AIR QUALITY

As discussed in the previous section, there have been many regulations on the federal, state, and local levels which have been implemented over the past several decades. On a national level, the average ozone design value has decreased 32 percent over the last few decades.

Statewide Ozone Trends since 1990

Over the past several decades, a decreasing trend in the both the annual statewide average concentration in the ozone design value, as well as average ozone levels, has been observed in every North Carolina region. While the year to year trend in ozone concentrations are variable, an overall decreasing trend is apparent. As discussed above, this decrease in concentration is likely due to the regulations passed by the state and local government agencies.

Regional Ozone Trends since 1990: Charlotte Region

As North Carolina's most populated region, the Charlotte area has had the hardest time attaining the National Ambient Air Quality Standards for ozone. Although the general trend in ozone concentrations does show a reduction in the concentration, the Charlotte area continuously observes higher concentrations than every other region. Two fairly significant reductions in ozone levels occurred after 2002 and then again after 2007. These two reductions are likely due to two major regulations implemented at the state and local level. As aforementioned, the Clean Smokestacks Act was enacted in 2002 at the state level while the GRADE program was implemented at the local level in 2007. However, these trends don't appear just in the year-to-year analysis. Looking at the trends in the three-year average concentrations of ozone from 1995 to 2015, the reduction in ozone concentrations in the early-to-mid-2000s is prominent. Comparing these averages with both standards, it is evident that this area frequently observed concentrations above both the 2008 ozone standard (75 ppb) and the 2015 ozone standard (70 ppb). However, there are two important things to note: there was a fairly significant reduction in the average ozone between the 2001–2003 period and the 2004–2006 period, and the 2013–2015 period saw average concentrations that were below the 2008 ozone standard (which was current, at the time) as well as just below the 2015 ozone standard.

These findings show that the ozone regulations released on the federal, state, and local level have positively impacted the region, not only on a year-to-year basis but on a more permanent basis as well.

Regional Ozone Trends since 1990: Triangle Region

Although the general concentrations of ozone observed were not as high as what was observed in the Charlotte region, the Triangle region did see higher concentrations than much of the state of North Carolina. While concentrations of ozone climbed during the 1990s, the implementation of the NC Clean Smokestack Act, among other major pieces of legislation, helped improve the region's air quality. Looking at the trends in the three-year average concentrations of ozone from 1995 to 2015, the reduction in the average ozone concentration between 2001 and 2003 and 2004 and 2006 is fairly extreme, on the order of ~15 ppb. Since then, there is a steadily decreasing trend in the three-year averages of ozone. However, when comparing these three year averages with the two standards (2008 NAAQS and 2015 NAAQS), it is clear that the average trends did not decrease such that they were less than the 2008 ozone standard of 75 ppb until the 2013–2015 period. However, the reductions in the average ozone concentration for that period reduced enough such that the period was also below the current (2015) ozone standard of 70 ppb.

Regional Ozone Trends since 1990: Triad Area

While the Triad region, in general, saw lower concentrations of ozone than both the Charlotte and Triangle regions, the concentrations were still elevated such that they exceeded the U.S. EPA ozone standards on a fairly frequent basis. As was seen elsewhere across the state, concentrations of ozone in this region rose steadily through the 1990s before peaking in the early 2000s and then dramatically reducing due to regulations placed on the emissions for precursor gases of ozone into the atmosphere. As previously mentioned, it is likely that this region saw the largest change in ozone due to the Clean Smokestack Act due to the presence of a large coal-burning power plant (Belews Creek Steam Station) in the region. As was seen in both the Charlotte and Triangle regions, the trends in the three-year average concentrations of ozone from 1995 to 2015 show the reduction in the average ozone concentration between 2001 and 2003 and 2004 and 2006. Since then, the three-year average concentrations of ozone stayed fairly constant (hovering around 80 ppb) before reducing into the upper 60 ppb range from 2013 to 2015. When comparing these three-year averages with the 2008 NAAQS and 2015 NAAQS, it is clear that it wasn't until 2013–2015 that this region saw an average three-year concentration of ozone below both the 2008 and 2015 ozone eight-hour standard.

CURRENT STATUS OF OZONE IN NORTH CAROLINA

Over the past few decades, ozone concentrations have significantly improved across the state of North Carolina. Thanks to a number of different important air pollution regulations set at the federal, state, and local levels on major precursor gases of ozone (such as NO_x), concentrations of ozone have steadily decreased such that nearly every region in the state, with the exception of the Charlotte region, is deemed in attainment for the 2008 NAAQS for ozone. However, the Charlotte region was reclassified as maintenance as of early 2016. While it is likely that some of North Carolina will go back into nonattainment for the new 2015 NAAQS for ozone, the reduction in the ozone levels that the state of North Carolina has made in the past decade is significant. This is in large part due to a combination of regulation and voluntary programs set at the federal, state, and local levels. North Carolina has reduced the average concentration of ozone by approximately 30 percent. For this air quality to be sustainable, emissions of ozone-forming pollutants must be kept at current levels, or further reduced if possible.

THE FUTURE FOR OZONE IN NORTH CAROLINA

The population and economy of North Carolina have grown substantially over the last 50 years, leading to increased energy usage and transport (both emit ozone precursor gases). Fuel combustion generally produces air pollutants that form urban smog, and in the 1970s, smog went along with urban growth in the state. Since then, though, the state has been mitigating the smog problem, and has been able to keep pace with increasingly stringent standards for air quality. This has been made possible by continuous improvements in air pollution controls, resulting in lower emissions per unit of production or per vehicle mile traveled. These controls have been implemented at the national, state, and local levels.

In the next century, the population growth and economic growth of the state are expected to continue to increase. In addition, formation of urban smog is expected to be enhanced by warming climate (Liu and Zhang 2013, 259–276). Standards for clean air may also continue to become more ambitious in the future. All of these factors mean that continuous reductions in emissions per unit of production and per mile traveled will be needed in order to maintain good air quality. As in the past, these reductions will depend on policies and actions at the national, state, and local levels.

Many initiatives are already under way to reduce future emissions of air pollution. For instance, U.S. EPA standards require improved mileage and reduced air pollution emissions for new cars and trucks. Federal and state policies also encourage the installation of renewable energy technologies, such as wind and solar energy generation, which eliminate smog-forming emissions. Local public transportation programs also reduce air pollutant emissions. Corporate programs and individual choices such as telecommuting and conservation measures can also affect emissions.

Since 2000, North Carolina has undergone sustained economic growth, while also maintaining high standards of air quality. In fact, the state has kept pace with reductions in the federal standards for urban smog. This sustainability can only be accomplished by continuing technological improvements to reduce the emission intensity per unit population and per unit of economic activity. This trend must continue into the future in order to maintain air quality with continued economic growth.

ACKNOWLEDGMENTS

We would like to acknowledge the NC DEQ Division of Air Quality and the Forsyth County Office of Environmental Assistance and Protection for providing the necessary data and generously assisting us in the research for this chapter. Support for this work was provided by the NASA Earth and Space Science Fellowship (NESSF) program, grant No. NNX15AN15H, and the Kenan Fund. We acknowledge North Carolina State University Air Quality Research Group. We would also like to thank Hofstra University, New York, for providing funding for one of the coauthors (VPA) to attend and participate in the Suburban Sustainability Symposium, LI, NY, during November 2016.

REFERENCES CITED

"Air Pollutant Emissions Trends Data, State Average Annual Emissions Trend, 1990–2014." 2016. https://www.epa.gov/air-emissions-inventories/air-pollutant-emissions-trends-data.

Aneja, Viney P., Gary T. Yoder, and S. Pal Arya. 1992. "Ozone in the Urban Southeastern United States." *Environmental Pollution* 75(1): 39–44.

"Annual Estimates of the Resident Population 2015 Population Estimates." 2016. https://factfinder.census.gov/faces/tableservices/jsf/pages/productview.xhtml?src=bkmk.

"Census of Population and Housing, 1960–2010." 2016. http://www.census.gov/prod/www/decennial.html.

Chameides, W. L., R. W. Lindsay, J. Richardson, and C. S. Kiang. 1988. "The Role of Biogenic Hydrocarbons in Urban Photochemical Smog: Atlanta as a Case Study." *Science* 241(4872): 1473–1475.

Civerolo, Kevin, Christian Hogrefe, Barry Lynn, Joyce Rosenthal, Jia-Yeong Ku, William Solecki, Jennifer Cox, Christopher Small, Cynthia Rosenzweig, and Richard Goldberg. 2007. "Estimating the Effects of Increased Urbanization on Surface Meteorology and Ozone Concentrations in the New York City Metropolitan Region." *Atmospheric Environment* 41(9): 1803–1818.

"Current Lists of Metropolitan and Micropolitan Statistical Areas and Delineations." 2016. https://www.census.gov/programs-surveys/metro-micro.html.

"Effects of Ozone Air Pollution on Plants." 2016. https://www.ars.usda.gov/southeast-area/raleigh-nc/plant-science-research/docs/climate-changeair-quality-laboratory/ozone-effects-on-plants/.

"Fact Sheet: Final Cross-State Air Pollution Rule Update for the 2008 NAAQS." 2016. https://www.epa.gov/sites/production/files/2017-06/documents/final_finalcsaprur_factsheet.pdf.

"Final Cross-State Air Pollution Rule Update." 2016. https://www.epa.gov/airmarkets/final-cross-state-air-pollution-rule-update#rule-summary.

"Green Book 1-Hour Ozone (1979) Area Information—NAAQS Revoked." 2018. Last modified December 21, 2018. https://www.epa.gov/green-book/green-book-1-hour-ozone-1979-area-information-naaqs-revoked.

Gschwandtner, Gerhard, Karin Gschwandtner, Kevin Eldridge, Charles Mann, and David Mobley. 1986. "Historic Emissions of Sulfur and Nitrogen Oxides in the United States From 1900 to 1980." *Journal of the Air Pollution Control Association* 36(2): 139–149.

"Highway Statistics, 1960–2014." 2016. http://www.fhwa.dot.gov/policyinformation/statistics.cfm.

Jacobson, Mark Z. 2012. *Air Pollution and Global Warming: History, Science, and Solutions.* New York: Cambridge University Press.

"Key Facts About the Clean Smokestacks Act." 2016. https://deq.nc.gov/about/divisions/air-quality/air-quality-outreach/news/clean-air-legislation/clean-smokestacks-act.

Liu, Xiao-Huan, and Yang Zhang. 2013. "Understanding of the Formation Mechanisms of Ozone and Particulate Matter at a Fine Scale over the Southeastern US: Process Analyses and Responses to Future-Year Emissions." *Atmospheric Environment* 74: 259–276.

1972–1995 Ambient Air Quality Trends Summary. 1998. Raleigh, NC: State of North Carolina, Division of Air Quality.

"North Carolina Counties with 8-Hour Ozone Violations 1997–1999." 2016. https://deq.nc.gov/about/divisions/air-quality/air-quality-data/data-archives-statistical-summaries/detailed-raw-ozone-data/north-carolina-counties-8-hour-ozone-violations-1997-1999.

"Ozone and Your health." 2009. https://www3.epa.gov/airnow/ozone-c.pdf.

"Ozone Trends." 2016. https://www.epa.gov/air-trends.

Redesignation Demonstration and Maintenance Plan for the Charlotte-Gastonia-Salisbury,

North Carolina 2008 8-Hour Ozone Marginal Nonattainment Area. 2015. Raleigh, NC: NC DAQ.

Section 110(a) Maintenance Plan for the 1997 8-Hour Ozone Standard for Greensboro/Winston-Salem/High Point 1-Hour Ozone Maintenance Area. 2013. Raleigh, NC: NC DAQ.

"State Energy Data System (SEDS): 2016–2014—North Carolina." 2016. https://www.eia.gov/state/seds/seds-data-complete.cfm?sid=NC#Consumption.

"Trends in Ozone Adjusted for Weather Conditions." 2016. https://www.epa.gov/air-trends/trends-ozone-adjusted-weather-conditions.

2001 Ambient Air Quality Report. 2008. Raleigh, NC: NC DAQ, 2008.01.

2005 Ambient Air Quality Report. 2010. Raleigh, NC: NC DAQ, 2010.01.

2011 Ambient Air Quality Report. 2013. Raleigh, NC: NC DAQ, 2013.01.

Wang, X. M., W. S. Lin, L. M. Yang, R. R. Deng, and H. Lin. 2007. "A Numerical Study of Influences of Urban Land-Use Change on Ozone Distribution over the Pearl River Delta Region, China." *Tellus B* 59(3): 633–641.

10

Skewed Sustainability and Environmental Injustice across Metropolitan St. Louis, Missouri

TROY D. ABEL, STACY CLAUSON, AND DEBRA SALAZAR

Urban and industrial air pollution exposure disparities continue to challenge the sustainability efforts of many North American cities. In this chapter, we review the case of the St. Louis region and its multidecade problems of racial segregation, inequitable development, and air pollution exposure inequality. In particular, we evaluate recent efforts by the St. Louis region to overcome past challenges and initiate a new regional plan that incorporates sustainability and livability principles through coordinated planning that integrates housing, land use, economic and workforce development, transportation, and infrastructure investments. This regional planning effort provides the opportunity to examine whether the sustainability-focused planning conducted by the region incorporates environmental equity issues, or whether air pollution disparities in the region continue to be a blind spot. We examine the adopted regional plan and apply measures of environmental inequality to examine the local distribution of pollution burden. Sustainability planning in St. Louis neglected air toxic disparities and their health inequities. A skewed sustainability perspective therefore obscures the region's persistent environmental and health injustices. Through our case study, we reveal the tensions between sustainability and livability planning and its response to issues of environmental injustice.

St. Louis: A Divided Region

Coined the "Gateway to the West" with a national park memorializing the launch of the Lewis and Clark expedition and westward expansion, St. Louis and its surrounding region illuminate sustainability's dilemmas. The city is located near the center of a bistate metropolitan area that includes five counties in Missouri and three in Illinois. Divided by the Mississippi and Missouri Rivers, the region is struggling to transform itself from struggling Rust Belt crossroads to a dynamic and resilient metropolis. The region faces sustainability challenges that will take decades to address. It ranks high in most indicators of racial disparity pertaining to employment, education, health, and poverty according to the East-West Gateway Council of Governments (EWGCOG 2015).

Discriminatory real estate and zoning practices, regional economic planning and urban renewal, and federal housing policies have all led to racial segregation, with African Americans isolated in declining communities (Heathcott and Murphy 2005). Tighe and Ganning (2015) observed that "race is the most predominant factor cleaving the city" (657). They documented decline and displacement in the city's predominantly Black and Northside neighborhoods, and a vibrant and stable development trajectory among its mostly White Southside communities. The region also remains one of the most hyper-segregated metropolitan regions by race in the United States (Landis 2019; Logan and Stults 2011). According to the strategic assessment comparing outcomes by race, Blacks have substantially higher rates of poverty, unemployment, and infant mortality while whites have household incomes that are twice those of Blacks. In addition, more than 13 percent of the region's poor reside in areas of concentrated poverty (EWGCOG 2013).

Other indicators point to significant sustainability challenges. Unemployment rates have not returned to pre-recession levels. Metropolitan St. Louis increased in population by 50 percent while its urbanized area sprawled by 400 percent. This sprawling suburban growth pattern has led to a mismatch of jobs and housing according to the United States Department of Housing and Urban Development (HUD 2012), with over 150,000 low- and moderate-income workers commuting more than 20 miles to work increasing pollution in the region and stretching household budgets. The region's particulate matter (PM) pollution was above the national standard and ranked among the worst places for asthma risk. In the past, the metropolitan area also failed to meet health-related air quality standards for ozone and lead while ambient air concentrations of cadmium were among the highest in the country according to the United States Environmental Protection Agency (EPA Region 5 1996, 20).

The St. Louis region's sustainability challenges are significant and span a range of issues. Next, we turn to the literature on sustainability planning to examine how the theories and practices associated with this approach are poised to address the challenges of environmental injustice.

Environmental Justice (EJ) and Sustainability

Sustainable development and environmental justice emerged as new environmental topics around the same time. Sustainable development became a widely recognized concept in the eighties when a United Nation's World Commission on Environment and Development (WCED 1987, 43) report defined it as "development that meets the needs of the present without compromising the ability of future generation to meet their own needs." Likewise, sociologist Robert Bullard (1983) introduced policymakers and social scientists to the distributional scrutiny of race and pollution that would become commonly known as environmental justice analysis. This research became a powerful force in the late eighties as advocacy groups and scholars started assessing the unequal distribution of environmental hazards. Appearing the same year as the Brundtland report, researchers with the United Church of Christ found poor and minority residents disproportionately concentrated in zip codes also hosting landfills (UCC 1987).

This and a few other studies spurred several national policy responses including an official EPA definition of EJ in 1998.

> The fair treatment and meaningful involvement of all people regardless of race, color, national origin, or income with respect to the development, implementation, and enforcement of environmental laws, regulations, and policies. Fair treatment means that no group of people, including racial, ethnic, or socioeconomic group should bear a disproportionate share of the negative environmental consequences resulting from industrial, municipal, and commercial operations or the execution of federal, state, local, and tribal programs and policies. (EPA 1998, 3–4)

For many, the concepts may seem intertwined, with EJ offering an ideal operationalization of the equity dimension of sustainability. Yet, a growing body of work has emerged to dispute this linkage (see Dobson 2003 among others). For example, some scholars note the dearth of research explicitly linking environmental justice and sustainability. "The two discourses and traditions of environmental justice and sustainability have developed in parallel and, although they have touched, there have been insufficient interpenetration of values, framings, ideas, and understandings" according to Agyeman, Bullard, and Evans (2003, 88). Another scholar described dissonance between the "Just Sustainability" paradigm and an "Environmentalist—Stewardship Sustainability" discourse (Agyeman 2005, 91). Others focus on the tension embedded within the concept of sustainable development itself.

For example, one planning scholar did not envision a Venn diagram with a harmonious intersection of economic development, ecological conservation, and social equity. "Planners have to redefine sustainability, since its current formulation romanticizes our sustainable past and is too vaguely holistic" (Campbell 1996, 296). Instead, he portrayed three conflicts that divide sustainability's three dimensions into a planner's triangle. First, a resource conflict pits economic growth and efficiency against environmental protection. The second division is a development conflict between social justice, economic opportunity, and income equality versus environmental protection. Third, a property conflict divides economic growth and efficiency from social justice, economic opportunity, and income equality.

Godschalk (2004) added a fourth corner to form what he called the Sustainability/Livability Prism. It illuminates three additional land use planning dilemmas: (1) a growth management conflict; (2) a green cities conflict; and (3) a gentrification conflict. The growth management divide pits livability against economic growth and those pushing for managing development versus those committed to unfettered market processes. The green cities conflict is a second tension between the development of natural and built environments. For example, the stewardship of natural resources like water quality can overshadow the quality of the built environment and its influence on public health. The St. Louis case illuminates some of these conflicts overlooked in most sustainability planning and the perspective that EJ analysis offers.

SUSTAINABILITY PLANNING IN ST. LOUIS

In 2010, the city, St. Louis County, and eight other organizations launched a planning effort with support from the U.S. Department of Housing and Urban Development (HUD 2010) Sustainable Communities Regional Planning Program (SCRPG). This federal grant opportunity signaled renewed federal support for regional planning that integrated housing and transportation decisions and incorporated livability, sustainability, and social equity values into land use plans and zoning (Marsh 2014). Researchers have deemed SCRPG a model of equity planning, addressing equity issues in both goal setting and implementation actions (Zapata and Bates 2016, 7). The effort stemmed from the Partnership for Sustainable Communities, linking HUD, the Department of Transportation (DOT), and the Environmental Protection Agency (EPA) to promote livability, emphasizing "transportation choice, greater housing options, and locally-driven strategies to protect natural resources while respecting existing communities" (Marsh 2014, 31).

St. Louis's efforts aim to "build an inclusive and opportunity-rich region" (OneSTL 2013, 4). A key outcome of this process was the first regional plan, OneSTL, a plan for a prosperous, healthy, vibrant St. Louis region, which was adopted by the East-West Gateway Council of Governments, the region's Metropolitan Planning Organization responsible for distributing transportation funding. OneSTL is based on the concept of sustainability, envisioned in the plan as a series of nested circles, situating the economy within society and social life embedded within the environment. As noted in OneSTL, "An effective plan for a sustainable future considers all three components as equally important in creating a prosperous, healthy, and vibrant future" (OneSTL 2013, 11).

The goals, objectives, implementation strategies and performance measures of OneSTL are organized around nine themes including commitments to reducing racial economic disparities, improving air quality, and improving the quality of life in low-income neighborhoods among others. The plan thus spans a broad range of issues, and though not a city planning policy, it provides a framework for the cities within the region to address many of the 38 elements included in Portney's Index of Taking Sustainable Cities Seriously (2013). Portney's Index was devised to gauge whether cities were taking sustainability seriously by implementing a range of sustainable policies and programs including household solid waste recycling, limited downtown parking, and air pollution reduction plans. St. Louis did not have the latter and ranked 46th out of 55 cities placing this city in the bottom quartile of urban sustainability efforts according to Portney (2013, 81). Below we show the importance of integrating environmental and social equity considerations in planning as inequality in exposure to air pollution remains a serious problem in the region.

MEASURING ENVIRONMENTAL INEQUALITY IN THE ST. LOUIS REGION

Metropolitan St. Louis is a highly segregated region, appearing on top-10 lists of regions throughout the United States with the highest measure of racial segregation, as gauged by the dissimilarity index (Logan and Stults 2011). In particular, the region is among several Midwestern regions that have persistent and very high black-white segregation. This level of segregation is likely to be associated with inequality in exposure to air toxics (Ard 2016).

Indeed, a 2008 study of the region described the region's industrial pollution as creating a highly skewed riskscape (Abel 2008). This study found that between 1987 and 2000, three facilities were responsible for one-quarter of the region's toxic air pollution emissions risk and were adjacent to East St. Louis. Ninety-eight percent of this city's population was African American, 25 percent was unemployed, and 49 percent lived below the poverty level (Abel 2008). Likewise, a more recent study found that in St. Louis, "census tracts with the highest levels of both racial isolation of Blacks and economic isolation of poverty were more likely to be located in air toxic hotspots than those with low combined racial and economic isolation" (Ekenga, Yeung, and Oka 2019, 1). Below we apply a different set of measures to assess inequality in exposure to air pollution during the period 2010–2014.

Drawing on recent developments in research on environmental inequality, we examine whether the patterns Abel (2008) observed persist. Using the EPA's Risk-Screening Environmental Indicators (RSEI) tool, we calculate three measures of environmental inequality: minority, African American, and poor shares of toxic concentrations following prior studies (Ash et al. 2013; Boyce, Zwickl, and Ash 2016; Salazar et al. 2019). These measures allow us to evaluate disproportionate exposure to point-source pollution emitted from large-scale industrial sources.

The RSEI tool uses chemical release information provided by large-scale industrial firms that are required to report their emissions under the Toxic Release Inventory (TRI) program (Figure 10.1). Specifically, we incorporated RSEI Geographic Aggregated Microdata (also known as RSEI-GM), which estimates how chemical releases from multiple TRI facilities combine and pose risk to residents in the surrounding community. RSEI-GM reports toxicity-weighted exposure concentrations, as well as the resulting human health hazard, aggregated over an 810-meter square grid covering the United States.

Our analysis uses the "ToxCon" variable in the RSEI-GM aggregate dataset, which provides a unitless indicator of exposure hazard that can be used to rank relative impacts (Ash and Fetter 2004; Downey et al. 2008; Schmidt 2003; US EPA 2016a). Following Salazar and her coauthors (2019) we first averaged RSEI toxic concentration values across multiple years (2010–2014)

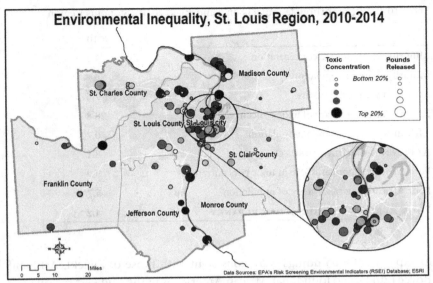

Figure 10.1. Industrial facilities reporting emissions under the Toxic Release Inventory (2010–2014).

in order to even out annual variation in emissions. We then aggregated grid cell data to U.S. Census blocks, using a crosswalk developed by EPA. We then summarized data to census tracts and weighted by the area of the tract to develop an area-weighted toxic concentration value for each census tract. This variable was then used, in combination with census data from the 2011–2015 American Community Survey, to generate the inequality measures.

Strikingly, minority discrepancy, developed by Ash and his coauthors (2013), reflects the percentage that exposure of minorities within the St. Louis region exceeds the minority population in the region (Table 10.1). The exposure does not indicate the level of risk, only the extent to which risk is disproportionately borne by minorities in the region.

Minorities in the region bear a disproportionate share of the toxic concentration, with a minority discrepancy of 7.6 percent (Table 10.1). The minority discrepancy would place the St. Louis region in the "Medium" ranking (75–95th percentile of 934 CBSAs throughout the United States) calculated by Ash and his colleagues (2013). Since African Americans are by far the largest racial minority in the St. Louis region we calculated an African American discrepancy, which shows that this group endures 8.1 percent more exposure to air pollutants than their share of the population. African Americans living below the poverty line also have a disproportionate share

Table 10.1. Disproportionate impacts on minorities (2010), St. Louis Region

	2010
Minority share of toxic concentration	34.3
Minority share of population	26.7
MINORITY DISCREPANCY	**7.6**
African American share of toxic concentration	28.8
African American share of population	20.7
AFRICAN AMERICAN DISCREPANCY	**8.1**
Poor African American share of toxic concentration	9.9
Poor African American share of population	5.7
POOR AFRICAN AMERICAN DISCREPANCY	**4.2**

of exposure. Our findings were consistent with those of Zwickl, Ash, and Boyce (2014) who found that African Americans in the Midwest region have higher average exposure than whites throughout their income range.

Drawing on research that shows that inequality is greatest in the most polluted places (Ard 2016; Collins, Munoz, and JaJa 2016), we also calculated exposure ratios between different population subgroups at the 90th percentiles. Adopting a method from Boyce and his coauthors (2016), we created an array of census tracts in ascending order of toxic concentration. Then each census tract's share of the region's minorities and poor families was calculated (as well its share of white and nonpoor families) and a running summation of the respective shares is ordered by toxic concentration. We identified census tracts that represent the 90th percentiles of exposure for each group and used the respective census tracts' toxic concentration to calculate the ratios. Ratios greater than one indicate patterned inequality in which the poor (or people of color in the race ratios) are disproportionately burdened by exposure to pollution (Salazar, Clauson, Abel, and Clauson 2019).

The results, taking seriously the positive skew in the distribution of pollution, focus on members of each demographic (e.g., whites) who endure the worst pollution (90th percentile exposure) for their group. Given this focus on the upper end of the distribution, highly exposed persons of color endure more serious pollution risk than highly exposed whites (Table 10.2). African Americans are the most burdened group. Further, poor, minority households endure more pollution than white households in the most polluted parts of the region.

Table 10.2. Disproportionate impacts on minorities and low-income at the 90th percentile of exposure (2010), St. Louis Region

	2010
RACE	
Ratio, minority to white exposure @ 90th percentile	1.45
Ratio, African American to White exposure @ 90th percentile	1.63
Ratio, Hispanic to White exposure @ 90th percentile	1.02
CLASS	
Ratio, poverty to nonpoverty exposure @ 90th percentile	1.79
RACE AND CLASS	
Ratio, poor minority to poor White @ 90th percentile	1.00
Ratio, poor minority to White @ 90th percentile	4.50

These results are consistent with Ard's (2016) findings regarding the co-occurrence of racial segregation and unequal exposure to pollution. Both on average and in the most polluted settings, racial minorities and the poor endure disproportionate industrial air pollution. To further assess the relation, we also conducted a spatial analysis of the exposure data to identify and locate potential clusters of high or low values of exposure. A local Moran's I spatial analysis was used to determine which census tracts within the St. Louis region are most similar and dissimilar in terms of exposure values. The local Moran's I calculation was used to test the significance of such clusters (Mitchell 2005). A Euclidean inverse-weighted, one nearest-neighbor minimum spatial weights matrix was generated in ArcGIS 10.3. A False Discovery Rate (FDR) correction, applied to control for multiple test and spatial dependency issues, was used to remove the weakest statistically significant p-values, based on an ordered list.

The output of the local Moran's I includes a z-score, which is used to generate a Moran scatterplot, with the following quadrants: (1) "high–high" clusters, where high values (such as high toxic concentration) are surrounded by other high-value tracts, (2) "low–low" clusters, where low values (such as low toxic concentration) are surrounded by other low-value tracts, (3) "high–low" outliers, where high values (such as high toxic concentration) are surrounded by low-value tracts, and (4)"low–high" outliers, where low values (such as low toxic concentration) are surrounded by high-value tracts (Figure 10.1). Tracts not found to have substantial spatial association with their neighboring tracts are deemed not significant. In this case, the results of the local Moran's I analysis identified high–high clusters where

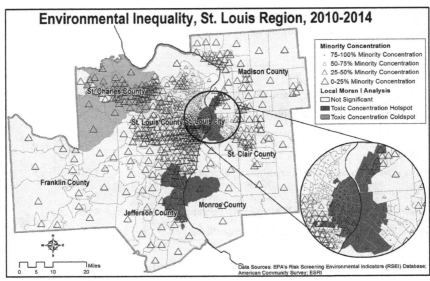

Figure 10.2. Moran typology of toxic concentration and race in St. Louis region (2010–2014).

high toxic concentration clustered with high minority concentrations, as well as low–low clusters, where low toxic concentration tracts clustered with low minority concentrations (Figure 10.2).

Our spatial analysis corroborates Abel's 2008 study and finds that environmental inequality is a persistent feature of the St. Louis region. A new environmental justice screening tool for the EPA confirms our findings. EJSCREEN was released in 2015 and we used this tool to compare six indexes in the high–high area of East St. Louis and the low–low area of St. Charles, Missouri, with a two-mile spatial buffer. The EPA data sources included (1) the National Air Toxics Assessment (NATA) Air Toxics Cancer Risk; (2) the NATA respiratory Hazard Index; (3) the NATA Diesel PM measure; (4) monitored $PM_{2.5}$ data; (5) monitored ozone data; and (6) traffic proximity and volume (EPA 2016).

An East St. Louis buffer exceeded the 90th percentile for the U.S., EPA Region 5, and the state for all six air pollution indices. For example, the NATA air toxics cancer risk results reached the 96th percentile in the state, the 98th percentile in the region, and the 91st percentile for the country. Conversely, the St. Charles buffer was well below the 50th percentile for all six indices. This area of the St. Louis region landed in the 38th, the 30th, and the 23rd percentiles for the state, region, and country on the NATA air toxics cancer risk. These results confirm both the high levels of pollution in parts of the region and the racially patterned inequality in exposure to pollutants.

SUSTAINABILITY'S BLIND SPOTS IN ST. LOUIS

With HUD's SCRPG, sustainability principles emerged as a key frame used to guide the development of regional plans. The St. Louis region engaged in a regional planning effort based on the promises of sustainable planning—by engaging the ecological, economic and equitable issues confronting the region. It is anticipated that the region would improve its capacity to maintain or provide a high-quality life for all residents. In defining what sustainability means for the St. Louis area, the regional plan states the following:

> Developing sustainably means developing and maintaining a strong economy, and it also means supporting the physical amenities of a healthy community—clean water and air, attractive and accessible recreation, supportive communities with good educational opportunities, safe streets and neighborhoods, and effective transportation services. (OneSTL 2013, 9)

While the plan has not been in place for sufficient time to evaluate its outcomes, the construction of the plan—and the issues addressed or not addressed within—reveal how sustainability planning, even when it includes social equity goals, overlooks how social and environmental factors are linked and must be considered together when framing the discussion around equity.

One glaring blind spot in the plan is the consideration of the spatial distribution of exposure to toxic air pollutants. Our analysis shows that there is unequal distribution of air toxics exposure, patterned by race and intertwined with the racial segregation patterns in the region. Further, the region's plans acknowledge that health outcomes are geographically concentrated and that race disparities exist (OneSTL 2013, 25). Yet, despite these social and environmental factors, there are no goals or implementation actions that seek to address these toxic air pollution inequities.

While the region addresses needed improvements to air quality, the focus is on emissions associated with transportation, targeting reductions in vehicle miles travelled. As stated in OneSTL, "the region's poor air quality is a contributing factor to St. Louis having the highest risk of asthma in the nation. Attention should be given to alleviating congestion and improving the region's air quality in an effort to promote public health and prosperity"

(26). Further, the plan does not address the variability in air quality that exists throughout the region. As Figure 10.2 demonstrates, there are both high and low outliers, where exposure levels differ significantly from the regional average. A focus on a regional average obscures the local variability that exists, and how this contributes to geographical disparities. It also allows planners to overlook small-scale changes that could generate incremental improvements. For example, OneSTL focuses on green infrastructure improvements to address stormwater and flooding concerns, but does not address the potential opportunities to address airshed issues at the local scale in much the same manner.

This blind spot is important, as without an understanding of how environmental and social factors relate, the plan overlooks how toxic air pollution burdens function as a component of the region's structural problems, operating alongside economic and social challenges. In St. Louis, residential segregation and income inequality continue to reinforce structural inequality, and these patterns are both influenced by and subsequently influence environmental inequality and disparities in health outcomes.

This blind spot was recognized by HUD in developing updated Affirmatively Furthering Fair Housing rules to improve implementation of the 1968 Fair Housing Act. As described by HUD, the new rules were intended to address historic patterns of segregation, promote fair housing choice, and foster inclusive communities. In adopting the new rule, HUD declared that despite over 40 years under the Fair Housing Act, "the ZIP code in which a child grows up all too often remains a strong predictor of that child's life course" (HUD 2015, 42348).

HUD's new rules now require consideration of the spatial distribution of environmental health hazards (24 CFR Part 574). HUD program participants are required to conduct an integrated assessment and planning process. To support this process, HUD developed an assessment tool that includes an Environmental Health Index. It summarizes potential exposure to harmful toxins at a neighborhood level and also utilizes NATA data similar to our analysis. The adoption of such measures signals the importance of considering the ways in which local, regional, and federal policies impact access to opportunity, and highlight how exposure and environmental health are key dimensions to include in an equity analysis, operating alongside other important opportunity indicators, such as poverty, education, employment, and transportation. Yet, the latest OneSTL planning performance report neglects to include similar metrics (EWGCOG 2015).

The region's plan has other issues of note that reveal the limitations in sustainability planning. First, while the region recognizes that uneven access to opportunity exists and should be addressed, the strategies the region employs to address these issues focus on providing housing options within opportunity-rich areas. Goals and policies addressing necessary improvements in neighborhoods lacking opportunity are underdeveloped and focus on redevelopment, a strategy that has led to displacement and gentrification in other regions (Immergluck 2009; Wyly and Hammel 1999).

Second, OneSTL, a blueprint plan, lacks detail and incorporates only voluntary measures. A HUD spotlight on the region noted that the plan was intended to be flexible, allowing local governments to customize elements to meet their needs. As stated by a planner from the region, "We are not trying to set up regionwide mandates from MPO [Metropolitan Planning Organization] to municipalities, but [instead] give them tools they can select from to figure out how to rally sustainability for their community. We're all going through the same future, but we may have a different path" (HUD 2012). This approach may make a regional plan more politically palatable for a divided region, but it limits its reach and potential to recognize and begin to ameliorate the structural inequalities that exist.

What St. Louis Tells Us about Sustainability

A renewed federal effort to support regional planning provided funding for the St. Louis region to identify a shared vision of a sustainable future, one that is inclusive and rich with opportunities. The plan that emerged from this effort took great strides toward documenting the unequal outcomes created through decades of discriminatory housing, employment, and planning practices and identifying new goals and policies aimed at closing the divides that remain in the region. OneSTL provides a framework for addressing sustainability's three Es (economy, ecology, and equity) and goes beyond Portney's Index, which gauges whether communities are taking sustainability seriously, by tackling social equity issues, including racial segregation, income inequality, and civic engagement issues. While OneSTL marks a significant departure from past planning initiatives, our work suggests several remaining limitations, which are illustrative of sustainability's persistent blind spots.

Our research reveals the uneven air pollution riskscape that exists within the region, linked with the patterns of inequitable development that have also contributed to housing segregation and racially patterned income inequality. The skewed riskscape is overlooked within OneSTL. This is just one example of the ways in which sustainability planning, even when it considers the three *Es*, fails to fully integrate them. As a result, sustainability as a practice and academic field misses environmental injustice.

Further, while the air pollution riskscape provides one lens through which to consider environmental inequality, an integrated analysis should extend beyond this to include multiple overlapping ways in which environmental inequality forms and is reconstituted (Pellow 2000). For sustainability to evolve, its theorists and practitioners need to engage in a more honest discussion about the tensions that exist among the three *Es*. Assessment of the spatial variation in outcomes and the underlying processes that have historically and continue to drive these patterns is an important first step, one that has been signaled by HUD in its new Affirmatively Furthering Fair Housing rules. While this is a promising development, it remains to be seen whether this will foster a productive dialogue on how to more fully integrate issues of environmental inequality into sustainability planning.

REFERENCES CITED

Abel, Troy D. 2008. "Skewed Riskscapes and Environmental Injustice: A Case Study of Metropolitan St. Louis." *Environmental Management* 42(2): 232–248.

Agyeman, J. 2005. *Sustainable Communities and the Challenge of Environmental Justice.* New York: New York University Press.

Agyeman, J., R. D. Bullard, and B. Evans, eds. 2003. *Just Sustainabilities: Development in an Unequal World.* Cambridge, MA: MIT Press.

Ard, Kerry. 2016. "By All Measures: An Examination of the Relationship Between Segregation and Health Risk from Air Pollution." *Population and Environment* 38(1): 1–20.

Ash, Michael, James K. Boyce, Grace Chang, and Helen Scharber. 2013. "Is Environmental Justice Good for White Folks? Industrial Air Toxics Exposure in Urban America." *Social Science Quarterly* 94(3): 616–36.

Ash, Michael, and T. Robert Fetter. 2004. "Who Lives on the Wrong Side of the Environmental Tracks? Evidence from the EPA's Risk-Screening Environmental Indicators Model." *Social Science Quarterly* 85(2): 441–462.

Boyce, James K., Klara Zwickl, and Michael Ash. 2016. "Measuring Environmental Inequality." *Ecological Economics* 124(2016): 114–123.

Bullard, Robert. 1983. "Solid Waste Sites and the Black Houston Community." *Sociological Inquiry* 53: 273–288.

Campbell, S. 1996. "Green Cities, Growing Cities, Just Cities? Urban Planning and the Contradictions of Sustainable Development." *Journal of the American Planning Association* 62(3): 296–312.

Collins, M. B., I. Munoz, J. JaJa. 2016. "Linking 'Toxic Outliers' to Environmental Justice Communities." *Environmental Research Letters* 11(1): 1–9.

Dobson, A. 2003. "Social Justice and Environmental Sustainability: Ne'er the Twain Shall

Meet." In *Just Sustainabilities: Development in an Unequal World*, edited by Julian Agyeman, Robert Doyle Bullard, and Bob Evans. Cambridge, MA: MIT Press, 83–98.

Downey, Liam, Summer Dubois, Brian Hawkins, and Michelle Walker. 2008. "Environmental Inequality in Metropolitan America." *Organization & Environment* 21(3): 270–294.

East-West Gateway Council of Governments (EWGCOG). 2013. "OneSTL: Many Communities, One Future." St. Louis, MO: East-West Gateway Council of Governments. https://www.ewgateway.org/wp-content/uploads/2017/08/OneSTL_FinalPlan-web.pdf

East-West Gateway Council of Governments (EWGCOG). 2015. "Where We Stand: The Strategic Assessment of the St. Louis Region." St. Louis, MO: East-West Gateway Council of Governments. https://www.ewgateway.org/wp-content/uploads/2017/07/WWS2015.pdf.

Ekenga, Christine C., Cheuk Yui Yeung, and Masayoshi Oka. 2019. "Cancer Risk from Air Toxics in Relation to Neighborhood Isolation and Sociodemographic Characteristics: A Spatial Analysis of the St. Louis Metropolitan Area, USA." *Environmental Research* 179: 108844.

Godschalk, D. R. 2004. "Land Use Planning Challenges: Coping with Conflicts in Visions of Sustainable Development and Livable Communities." *Journal of the American Planning Association* 70(1): 5–13.

Heathcott, Joseph, and Marie Agnes Murphy. 2005. "Corridors of Flight, Zones of Renewal Industry, Planning, and Policy in the Making of Metropolitan St. Louis, 1940–1980." *Journal of Urban History* 31(2): 151–189. DOI: 10.1177/0096144204270715.

Immergluck, D. 2009. "Large Redevelopment Initiatives, Housing Values and Gentrification: The Case of the Atlanta Beltline." *Urban Studies* 46(8): 1723–1745.

Landis, John D. 2019. "Black-White and Hispanic Segregation Magnitudes and Trends from the 2016 American Community Survey." *Cityscape* 21(1): 63–86.

Logan, John R., and Brian J. Stults. 2011. "The Persistence of Segregation in the Metropolis: New Findings from the 2010 Census." Census Brief prepared for Project US2010. http://www.s4.brown.edu/us2010/Data/Report/report2.pdf.

Marsh, Dwayne S. 2014. "The Sustainable Communities Initiative: Collective Impact in Practice." *Community Investments* 26(1): 30–36.

Mitchell, Andy. 2005. *The ESRI Guide to GIS Analysis Volume 2: Spatial Measurements and Statistics*. Redlands, CA: ESRI.

Pellow, David N. 2000. "Environmental Inequality Formation: Toward a Theory of Environmental Injustice." *American Behavioral Scientist* 43(4): 581–600.

Portney, Kent. 2013. *Taking Sustainable Cities Seriously: Economic Development, the Environment, and Quality of Life in American Cities*. 2nd ed. Cambridge, MA: MIT Press.

Salazar, Debra J., Stacy Clauson, Troy D. Abel, and Aran Clauson. 2019. "Race, Income, and Environmental Inequality in the U.S. States, 1990–2014." *Social Science Quarterly* 100(3): 592–603.

Schmidt, Charles W. 2003. "The Risk Where You Live." *Environmental Health Perspectives* 111(7): A404–A407.

Tighe, J. R., and J. P. Ganning. 2015. "The Divergent City: Unequal and Uneven Development in St. Louis." *Urban Geography* 36(5): 654–673.

United Church of Christ (UCC). 1987. *Toxic Wastes and Race in the United States : A Na-*

tional Report on the Racial and Socio-Economic Characteristics of Communities with Hazardous Waste Sites. New York: United Church of Christ.

United States Department of Housing and Urban Development (HUD). 2010. "HUD Awards Nearly $100 Million in New Grants to Promote Smarter and Sustainable Planning for Jobs and Economic Growth." Last accessed January 31, 2020. https://archives.hud.gov/news/2010/pr10-233.cfm.

United States Department of Housing and Urban Development (HUD). 2012. "Grantee Spotlight: St. Louis Plans for Regional Renewal | HUD USER." *Sustainable Communities ENews.* June. https://archives.huduser.gov/scrc/sustainability/newsletter_052912_4.html.

United States Department of Housing and Urban Development (HUD). 2015. "Affirmatively Furthering Fair Housing; Final Rule." 80 *Federal Register* 136. Last accessed December 15, 2017. http://www.gpo.gov/fdsys/pkg/FR-2015-07-16/pdf/2015-17032.pdf.

United States Environmental Protection Agency (EPA) Region 5. 1996. *Agenda for Action.* EPA 905-K-96-002. Chicago, IL.

United States Environmental Protection Agency (EPA). 1998. *Final Guidance for Incorporating Environmental Justice Concerns in EPA's NEPA Compliance Analysis.* Last accessed January 30, 2020. https://www.epa.gov/sites/production/files/2014-08/documents/ej_guidance_nepa_epa0498.pdf.

United States Environmental Protection Agency (EPA). 2016a. "RSEI and NATA." Data and Tools. *US EPA.* May 12. Last accessed December 15, 2017. https://www.epa.gov/rsei/rsei-and-nata.

United States Environmental Protection Agency (EPA). 2016b. "EJSCREEN: Environmental Justice Screening and Mapping Tool." Last accessed June 16, 2017. https://www.epa.gov/ejscreen.

United States Environmental Protection Agency (EPA). 2016. "RSEI and NATA." Data and Tools. *US EPA.* May 12. Last accessed December 15, 2017. https://www.epa.gov/rsei/rsei-and-nata.

World Commission on Environment and Development (WCED). 1987. *Our Common Future.* Oxford: Oxford University Press.

Wyly, E. K., and D. J. Hammel. 1999. "Islands of Decay in Seas of Renewal: Housing Policy and the Resurgence of Gentrification." *Housing Policy Debate* 10(4): 711–771.

Zapata, Marisa A., and Lisa K. Bates. 2017. "Equity Planning or Equitable Opportunities? The Construction of Equity in the HUD Sustainable Communities Regional Planning Grants." *Journal of Planning Education and Research* 37(4): 411–424.

Zwickl, Klara, Michael Ash, and James K. Boyce. 2014. "Regional Variation in Environmental Inequality: Industrial Air Toxics Exposure in U.S. Cities." *Ecological Economics* 107: 494–509. DOI: 10.1016/j.ecolecon.2014.09.013.

11

Suburban Sustainability Governance in the Los Angeles Region, California

ELIZABETH MATTIUZZI

Environmental sustainability in planning typically brings to mind images of big cities—bike share systems, subways, and LEED-certified office buildings. Many of the major urban areas in the U.S. have the benefit of a prewar grid that lends itself to walkable development, along with the associated quality of life benefits and reduced vehicle miles traveled (VMT). Central cities have high land values, making infill development attractive. Suburbs, on the other hand, often have smaller populations per jurisdiction, a smaller public sector with fewer resources, and more dispersed land uses. Many suburban cities, towns, and unincorporated areas, although they might have a small historic grid, lack walkable neighborhoods.

Suburbs typically do not have the unified governance structure or leadership of a big city relative to their population size. Often the farther they are from the urban core, the lower their land values, making low-density development the most feasible land use. Suburbs are often in competition with one another, as well as with larger cities, for funding to implement sustainability goals such as smart growth, energy efficiency, and greenhouse gas (GHG) emission reduction. Yet low-density suburbs are the fastest growing areas in terms of population, and in many metropolitan regions they are the most affordable place to live (Kolko 2017). Suburban housing is often newer and more energy efficient per square foot, yet suburban households consume more energy per capita due to the larger size of homes than in urban areas (Estiri 2016, 1921).

This chapter explores the role that collaboration across jurisdictional boundaries plays in how suburbs carry out state and regional sustainability policies. Scholarship on suburban revitalization acknowledges the importance of governance and jurisdictional boundaries in determining policy and the distribution of resources (Hanlon 2009, 35; Dunham-Jones and Williamson 2011, 2–14), yet there is little explicit work on governance for sustainability among suburbs. Regional governance is a key part of achieving suburban sustainability, but the priorities of larger cities in a region often shape it. The politics and resources of a smaller, less central city or town can determine whether sustainability goals take root there. Depending on their priorities, suburbs can try to block state and regional sustainability agendas, cooperate with them, or band together to advance sustainability in a suburban context.

Three Southern California suburban areas illustrate the unevenness of the implementation of the state's climate framework. In the absence of federal intervention or a strong state bureaucracy for implementing climate policy at the local level, the mechanisms by which localities address sustainability are important to understand. Yet in the largest region in an environmentally-forward-looking state, the arrangements for implementing state climate goals remain relatively informal. Surprisingly, a patchwork of state and regional carrots and sticks has fostered collaboration on climate planning at the often-overlooked county and subcounty scales. The motivations for local jurisdictions to collaborate include avoiding being sued by the state, attracting state and regional grant dollars, and achieving efficiencies in environmental reporting requirements. Suburban jurisdictions do not necessarily have the capacity to engage in climate planning on their own, and they can avoid reinventing the wheel by working with county and subcounty agencies/organizations. The variety of approaches and degrees of success achieved by these Southern California suburbs in conforming to state climate goals has implications for how we understand regionalism, multiscalar governance, and sustainability policy in a suburban context.

Southern California Sustainability

The Los Angeles region encompasses six counties in southern California and nearly half the state's population. It provides an interesting case study in suburban sustainability, both because of its large number of diverse jurisdictions and counties in one sprawling region, and because it has been the site of much experimentation in sustainability at the subregional level.[1] A "subregion" is a group of jurisdictions, either within a county or encompassing an entire county, that are represented by some kind of governing body. "Subregional" agencies serve many of the nearly 200 jurisdictions in the Los Angeles region. In recent years, these subregional agencies have promoted suburban sustainability in response to mandates and incentives at the regional and state levels using different approaches and with different levels of success.

The sheer size of the Los Angeles region, the second largest urbanized area in the country, makes it an extreme case. Yet it is also a laboratory for understanding how smaller subsections of the region address suburban sustainability within the same regional and state policy environment. In this chapter, three examples of subregional sustainability governance illustrate different types of partnerships between different jurisdictions, the way they collaborate with one another and negotiate the regional and state frameworks for addressing sustainability in California, and their inventiveness in promoting private sector involvement in suburban sustainability.

In different regions of the U.S., different types of government and quasi-government agencies, such as planning districts, economic development agencies, associations of governments, and the like, serve the jurisdictions in a particular geographic area. In southern California, councils of governments (COGs) and county transportation authorities (CTAs) are two types of agencies that promote regional governance. Since they serve a population and a geographic area comparable to entire metropolitan areas in other parts of the country, these subregional agencies within the LA metro could provide lessons to these other regions.

COGs are voluntary organizations that represent the interests of cities and towns, particularly suburban jurisdictions (Wikstrom 1977, 130). They provide their members with research, technical assistance, a forum to learn from neighboring jurisdictions, and a way to respond with a unified voice to regional and state policy. They have the ability to apply for grants and help their member jurisdictions take advantage of incentives and address mandates from the state and region on sustainability and other issues. CTAs can play a similar role, but they came into existence to administer local transportation sales tax revenue as responsibility for transportation planning and management devolved to the county level in California in the 1980s and 1990s (Wachs 2003, 9–15; Elkind 2014, 176–188).

The Western Riverside COG (WRCOG), the San Bernardino Associated Governments (SANBAG), a CTA with COG functions, and the South Bay Cities, a COG in Los Angeles County, are examples of subregional agencies that coordinate actions by suburban jurisdictions on energy, land use, and transportation in suburban parts of the Los Angeles region. They have varied funding sources and levels of engagement in regional and local climate planning. They illustrate different approaches to subregional governance and different levels of success with promoting sustainability in suburban jurisdictions. However, they are for the most part responding to the same state and regional policies and incentives. In the context of the shared interests of their member jurisdictions, they engage in governance in a complex policy environment to leverage state and regional resources and reduce the burden of addressing sustainability issues on individual suburbs.

State Climate Policy in California

California is known for its sprawling suburbs and its traffic-induced smog perhaps as much as it is for its forward-thinking environmental policies. In 2006, the Governator, Arnold Schwarzenegger, signed landmark climate change mitigation legislation. Assembly Bill 32 (Pavley and Nunez 2006) regulates emissions of heat-trapping greenhouse gasses from mobile and stationary sources. It encompasses the supply and demand sides of emissions by regulating fuels, energy generation, and the efficiency of buildings, industry, and vehicles. Subsequent bills outlined the details of its implementation.

In 2007, the legislature followed up with Senate Bill 97 (Dutton 2007), requiring that environmental impact reports (EIRs),[2] in compliance with the California Environmental Quality Act (CEQA), consider greenhouse gas emissions as an environmental impact and describe how they will be mitigated. This law meant that local general plans[3] subsequently had to inventory GHGs. The significance of this was that now all counties, cities, and towns had to think about the impact of all of their activities and the development they allow on climate change.

The following year, Senate Bill 375 (Steinberg 2008) introduced a framework for addressing the contribution of land use-induced vehicle travel demand to state GHG emissions. The framework that SB 375 created relies on local jurisdictions to implement regional sustainability plans to promote compact development and needed housing production near public transportation. These plans, called Sustainable Communities Strategies (SCSs) are developed by metropolitan planning organizations (MPOs), the regional transportation agencies.[4] The law's central paradox is the tension between the state's strong local land use control and the requirement that cities zone for their MPO-allocated state housing targets. Ultimately, subregional agencies like CTAs have played an outsized role in helping cities, especially suburbs, benefit from incentives to implement SB 375 while protecting local authority over land use (Mattiuzzi 2016).

State and Regional Incentives for Local Sustainability

Incentives for sustainability measures, in the form of grant funds at the state and regional scales, assist local jurisdictions, and suburban jurisdictions typically take advantage of them collectively through subregional agencies. To help local jurisdictions implement California's climate framework, the state set up a Strategic Growth Council (SGC) within the Governor's Office of Planning and Research, the office that issues guidance on planning law to cities. The state's market-driven "Cap and Trade" auctions fund much of the grants that the SGC distributes to cities, regional MPOs, and subregional agencies representing multiple jurisdictions to support sustainable housing and community development. SGC grants are a key state incentive for local (and subregional) sustainability efforts, particularly around land use.

Local governments play a key role in energy efficiency in the built environment, both in terms of their own facilities and their residents. Southern California Edison (Edison) is one example of an investor-owned utility in the state that promotes energy efficiency in the context of the state's "decoupling" of energy profits from energy usage in the 1980s. Edison grants have helped subregional agencies coordinate planning for energy efficiency among suburban jurisdictions in the Los Angeles region.

The MPO in the LA region, the Southern California Association of Governments (SCAG), has a sustainability grant that provides an incentive for cities, counties, CTAs, and subregional COGs to do sustainability work. Examples of projects include corridor redevelopment, downtown revitalization, and station area planning. The grants are a source of cooperation between SCAG and local and subregional agencies and help localities start projects, leverage other sources of funding, and create examples of smart growth in Southern California.

The LA Region: Suburban Sustainability in Context

Population growth in California has greatly outpaced new housing supply, driving burdensome housing costs and commutes. In 2015, home values were 2.5 times the national average and rents were 1.5 times the national average (LAO 2015). The coastal regions of the state have experienced the greatest job growth and built the least housing, leading to an imbalance in the location of jobs and housing and long commutes from more affordable inland suburbs.

Average commutes in California's coastal metropolitan regions top an hour, well above the national average (Rapino and Fields 2013). Commuters from the "Inland Empire," the far eastern suburbs of San Bernardino and Riverside Counties, to Los Angeles County make up the largest share of megacommuters (over 90 minutes and over 50 miles) in the country (Rapino and Fields 2013). Coastal Orange County is close behind as a commute destination with high housing prices.

Subregional agencies in the Los Angeles region piece together different planning grants, including the SGC, Edison, and SCAG sustainability grants. These sustainability grants encourage these agencies to share their work with other jurisdictions, and efforts in one subregion are often replicated in other areas. SANBAG, WRCOG, and the South Bay COG are examples of subregional agencies with limited political power that have leveraged incentives to help suburban jurisdictions respond to state and regional sustainability mandates and incentives.

San Bernardino: Suburban Climate Planning in Response to State Action

After the passage of SB 97 in 2007 instituted the requirement that EIRs consider GHGs as a pollutant, then Attorney General Jerry Brown made an example of San Bernardino County with a lawsuit directed at the county-wide general plan EIR. A court ruled that the county's general plan did not meet the new standard for inventorying and mitigating GHG emissions. The message was clear: sustainability and emissions reductions are not just for larger cities, they are a responsibility of all jurisdictions in the state, even those far from the urban center of the region.

The county responded with a new climate action plan inventorying emissions for unincorporated areas and the activities of county government (Settlement Agreement 2007). But this left small suburban jurisdictions to figure out how to do the work of documenting their own GHG emissions. The situation prompted the suburban jurisdictions in San Bernardino County to work together via their CTA-COG, the San Bernardino Associated Governments (SANBAG). SANBAG's countywide climate action plan provided a template for local jurisdictions to make emissions reductions, and it provided a model for neighboring counties.

SANBAG prepared a countywide GHG inventory and climate action plan (CAP), with 21 out of 24 cities in the county choosing to participate. After calculating how much state actions would actually reduce emissions in San Bernardino County, SANBAG went about preparing a chapter of the CAP for each jurisdiction, with their input, that cities could adopt essentially whole cloth at the local level. SANBAG assisted cities with setting targets and selecting measures for reducing their emissions, and provided them with a tool for monitoring and tracking implementation of GHG reduction measures. A new bus rapid transit line, energy efficiency measures, and new bike-pedestrian infrastructure are among the planned actions.

A major motivating factor for jurisdictions to help shape and adopt their chapter of SANBAG's CAP was the potential cost savings of future environmental analysis. In addition to avoiding litigation by meeting the state mandate and savings on the GHG inventory itself, suburbs in San Bernardino County saw an opportunity for "tiering." Using the analogy of different tiers of environmental impact documentation, a lower-level project or policy (geographically or bureaucratically) can invoke part of a higher-level agency or plan's EIR without repeating the analysis (CEQA n.d.). By collaborating via a subregional agency, suburban jurisdictions attempted to increase their ability to take advantage of tiering benefits.

One example of how tiering would work for GHG emission reporting in suburbs in San Bernardino County would be if a local jurisdiction adopted language in the EIR for their general plan from the SANBAG countywide climate plan. Since the SANBAG language is consistent with the environmental review document for the SB 375–mandated Sustainable Communities Strategy in the Los Angeles region, a local jurisdiction can make their plans and policies consistent with those higher-level documents without doing additional analysis. In turn, new developments in a particular jurisdiction can use the same language on GHGs. According to one official, the project GHG screening tool that SANBAG developed with cities

> allows the developer to pick and choose from the menu of options, they can earn points, [and] if they earn a certain amount of points, the planner can say "you meet the level for being consistent with the regional plan." The idea is the developers won't have to include GHGs in their project-specific EIR. The GHG portion for a development study runs anywhere from $5,000 to $30,000. It's a cost-saving measure for the cities. . . . If cities achieve a certain amount of points . . . then the new development is considered consistent with the regional plan. (author interview)

The coordinated sustainability plan that SANBAG did for jurisdictions in San Bernardino County allows suburbs that might not otherwise have the resources to do so to make their local plans consistent with SCAG's SCS.

Through subregional collaboration, small jurisdictions, such as those in the Inland Empire, can coordinate their development with the climate change goals of the entire Los Angeles region.

Working together on a CAP has increased cooperation on sustainability issues across jurisdictions in San Bernardino County. According to one SANBAG official, prior to this process, jurisdictions

> were working in silos and just not aware of what others were doing. . . . Maybe the water districts were talking a little before, but not all of these water agencies have framed a comprehensive plan for water . . . [the development of which] was actually one of the first significant milestones [of CAP implementation]. (author interview)

Developing the CAP improved governance, or cross-jurisdictional cooperation, learning, and relationships, on the issue of water resources, where one city's actions might affect another.

The countywide CAP also helped lay the groundwork to attract funding for active transportation. In 2014, San Bernardino County received a combined $23 million for bicycle and pedestrian infrastructure and planning from a statewide competitive grant and from SCAG. According to SANBAG officials, this

> has never happened in our county [previously on this scale]. We would have never dreamed that our county would get any type of funding related to bike and ped[estrian] projects. We were successful because of all of the emphasis we put in active transportation [in the CAP] and all of the programs related to active transportation that our cities are on board with. (author interview)

Local officials anticipate spillover benefits from these investments in health, safety, and quality of life.

The state GHG inventory mandate and environmental review benefits helped motivate suburbs in San Bernardino County to address sustainability issues, including land use, energy and water efficiency, and bike-pedestrian infrastructure. The groundwork laid by the CAP gave local water agencies an opportunity to increase their coordination and helped attract bike-pedestrian infrastructure grants to a subregion that was not previously competitive for this funding. The coordination of a subregional agency helped make these changes in suburban sustainability possible.

RIVERSIDE COUNTY: INTEGRATED SUBURBAN CLIMATE ACTION PLANNING

The Western Riverside Council of Governments (WRCOG) is unusual for a subregional COG in the Los Angeles Region. In response to the pressure on their local transportation system created by new suburban development in different jurisdictions, their member cities came together in the early 2000s to agree on a voter-approved fee system for mitigating new development. Remarkably, jurisdictions agreed to a system where funding from their jurisdiction, some of which returns to them, also supports activities in other parts of Western Riverside. The Transportation Uniform Mitigation Fee (TUMF), modeled on a similar program started by the Coachella Valley COG, is leveed on all jurisdictions in western Riverside County but is mainly returned to a geographic area of the subregion that is composed of two to three cities.[5] It also supports subregion-wide projects and WRCOG operations. The system is robust enough that when a jurisdiction stopped paying, WRCOG won a settlement against them (Farren 2017).

State and regional sustainability grants, and the COG's role in energy efficiency, supported integrated climate action planning in Western Riverside. In 2010, WRCOG received an Edison grant to help jurisdictions prepare energy efficiency plans, which became part of the subregional CAP. WRCOG received one SGC grant to assist cities with preparing emissions inventories, forecasting emissions, and establishing locally-agreed-upon reduction targets, and a second grant to help them incorporate the GHG targets into their local plans. A SCAG Sustainability Grant also supported cities' incorporation of the subregional GHG goals into local land use plans.

Significantly, rather than compiling a chapter for each participating city, WRCOG prepared a single, integrated CAP with sections on different issues that all of the participating cities helped develop.[6] Cities agreed to individual emission reduction targets and reduction measures that cut across cities within the subregion. A SCAG sustainability grant helped WRCOG develop "model code amendments, ordinances, and general plan measures to implement measures in the CAP. Cities can take these template ordnances or code amendments and tailor them to their cities," according to one official (author interview). The officials responsible for sustainability in these suburban jurisdictions had worked together enough through the subregional COG that they saw themselves as having common goals.

Surprisingly, the planning process for the WRCOG climate action plan encouraged a race to the top among cities, rather than a race to the exits.

Like homeowners receiving a report on whether their electricity use is above or below average for their street, cities looked to their neighbors for a sense of what reasonable reductions were, and perhaps to be seen as pulling their weight. The participating jurisdictions agreed on an overall 15-percent GHG reduction target for 2020 below a 2010 baseline (WRCOG 2014). This includes state and regional activities that will affect the subregion, as well as local actions.

After meeting with the cities multiple times to gather their input and figure out a feasible level at which cities could participate in the CAP, COG officials realized they needed greater reductions to meet their target: "When we started, we met one-on-one with the cities multiple times . . . [to ask] 'what do you think is feasible, at what level can you participate, gold, silver, or platinum? What is politically feasible and technically feasible within your city?'" (author interview). After helping each city quantify their emissions, the COG realized they were just short of their target and needed cities to push for greater reductions.

The COG hesitated to show cities how much each of them was doing to reduce emissions, lest it cause them to want to reduce their efforts, as not to bear more of a burden than their neighbors. Yet, said one official, "sure enough, at the next meeting we showed them the charts and it started becoming like a friendly competition. . . . Everybody kind of challenged one another because no one wanted to be the city that was slacking even though there wasn't a penalty" (author interview). The most popular measures were for active transportation, particularly bike infrastructure. Examples include traffic signal coordination, reduced parking requirements for new development, and bike infrastructure. Like in San Bernardino County, the CEQA benefits of having local plans be consistent with a subregional CAP helped motivate suburban jurisdictions in Riverside County to participate in subregional GHG emission reduction efforts.

South Bay Cities: Subregional Government Coordination of Shared Mobility

The South Bay Cities Council of Government (South Bay COG) represents a group of hilly coastal and flat inland suburban jurisdictions in southwest Los Angeles County. The subregion's geography creates challenges for meeting state and regional sustainability goals and for taking advantage of incentives for smart growth. The South Bay does not have the combined urban density and high land values that make building a robust transit system or utilizing state infill incentives feasible for its neighboring cities, Los Angeles and Long Beach. Nor does it have the greenfield land (or planned transit stations with developable land around them) that might make CEQA tiering attractive for planning and building smart growth in the Inland Empire. The South Bay has large swaths of single-family neighborhoods and low-density commercial along arterial roadways, with higher home values in the hilly areas near the coast and lower values in the flats. These suburbs are what many consider "built out," although they have potential for redevelopment in commercial areas, as well as densification through accessory dwelling units.

Despite these disadvantages, the suburban cities in the South Bay have engaged in subregional sustainability governance through their COG. The South Bay COG has used its convening role and staff resources to develop a sustainability planning blueprint and pilot shared mobility programs. Its funding source is its member fees, making its revenue more limited than a COG or CTA with developer fees or sales tax revenue. This has put the South Bay COG in the position of needing to be entrepreneurial about leveraging the resources it has.

The South Bay has also benefited from some of the same regional and state incentives for sustainability as San Bernardino and Riverside. The COG received an Edison utility grant for a GHG emissions inventory and mitigation plan for the subregion and its cities. SGC funding supported the COG's development of the transportation, land and use, and energy chapters of its CAP. A state SGC grant also supported the COG's assistance to cities with developing their own CAPs and establishing criteria for determining a development project's consistency with that local CAP to take advantage of tiering benefits.

Subregional governance in the South Bay has had a unique focus on mobility and public-private partnerships to promote it. The South Bay's residential densities are not far behind more central areas of LA County, yet cities in the South Bay with postwar suburban infrastructure face obstacles to promoting low-carbon travel. "Measure R [LA County's transportation sales tax] really charted where the [transit] money is going for the next 30 years, which also solidified the fact that we're not going to have rail here," said one South Bay official (author interview). Regional competition from cities that have larger planning staffs and more existing civil society activity around smart growth make it harder to attract transit funds away from the center of the region. Furthermore, "nobody comes to the South Bay and sells their car. . . . A lot of our bus services are hourly or half-hourly and some don't run on weekends," according to a South Bay planner (author interview). The COG has worked to think outside of the chicken-and-egg problem of attracting transit and the densities to support it by focusing on vehicle sharing services and electric vehicles to reduce GHG emissions from travel and increase mobility for residents.

The Sustainable South Bay Strategy (Strategy) is a blueprint developed by the COG that tries to address transportation emission reduction in the context of a suburban area with limited transit. It encourages cities to construct suburban neighborhood centers at the intersection of arterial corridors with gradually decreasing density. The neighborhood centers would include high density commercial-residential mixed use, with medium density residential between nodes (Siembab et al. 2009). While in the urban core mixed use might refer primarily to mixing uses vertically within a single building, for suburban areas like the South Bay, horizontal mixed use can be more practical, with housing, particularly multifamily, next to or within walking distance of commercial uses (Chapple 2015, 209). In the South Bay and mature suburbs like it, a retail customer base must include people driving and people living within walking or biking distance (Boarnet et al. 2011, 129–159). The Strategy seeks to increase walking and biking within one to two miles of a neighborhood center, particularly for people living along the corridor, as well as to encourage people to use electric vehicles for three-to-five-mile trips to schools and shopping centers.

The Strategy aims to reduce vehicle trips, promote electric vehicle use, and build density along corridors to attract future transit options that the area's density does not currently support. Trips under three miles make up most of the trips other than the commute in the South Bay, according to the COG's research. As one South Bay COG representative put it, "Those trips are too short for transit and too long to walk. Nobody stands on a corner to wait for a bus to take them two miles" (author interview). While people are not likely to ditch their primary vehicle for commuting, they might use a different vehicle or mode to get around their neighborhood. Although implementing the Strategy's land use ideas would require further action from cities and developers, the South Bay COG coordinated two electric vehicle programs to promote its mobility goals.

Community electric vehicles and commercial carsharing might help bridge the gap between the long-term goal of densification of land use along corridors and the short-term goal of promoting mobility in suburbs. The South Bay COG piloted a neighborhood electric vehicle (NEV) program with support from the South Coast Air Quality Management District. NEVs are roughly the scale and power of a golf cart, which works well in the flat areas of the South Bay. Under current laws, NEVs can cross but not travel on higher-speed arterials. NEV sharing has shown promise as a way to promote senior mobility and reduced GHG emissions (Shaheen, Cano, and Camel 2016). The success of the NEV program motivated the South Bay COG to increase the reach and impact of vehicle sharing.

The South Bay COG coordinated a pilot program for commercial car-sharing as an emission reduction strategy. The South Bay COG used its existing relationships with multiple suburban municipalities to help negotiate the details of the program, including parking fees collected by cities, making the move feasible for the company. Car2go charges by the minute for the use of its vehicles, allows users to leave the vehicle anywhere within its service area, and pays cities in advance for the use of metered or permitted parking spaces. The parking issue required negotiation with cities for the company to pay in bulk for short-term and long-term parking spaces, and the COG assisted with this negotiation. Some cities experienced pushback from residents who were concerned about parking availability, but seven agreed to a one-year pilot.

The South Bay's CAP planning and vehicle sharing work provides a case of collaborative leadership by a subregional CAP. Unfortunately, Car2go decided to end the program after the one-year pilot (Dryden 2015). The company reported that their users, who numbered about 900, wanted to be able to take the vehicles outside the service area to the larger cities such as Los Angeles (Car2go 2015). This suggests not that the program was a failure, but that it helped identify a need for coordination, possibly by a regional agency like SCAG, across a greater swath of the Los Angeles region.

A key innovation for suburban sustainability in the South Bay was the role of the COG in coordinating both the NEV program and a carsharing pilot between cities. On a variety of sustainability issues, one role of a subregional COG could be to coordinate between small, suburban jurisdictions and private sector partners. As one COG representative noted:

> If you can create a model permitting process or something around electric vehicle chargers or . . . solar, that can also . . . [make] it easier for developers if they know that the rules of the game are consistent

and they are not just figuring out a different system for every city ... because a lot of contractors and developers ... are going to work in the whole [subregion]. (author interview)

A COG can leverage its staff time, its ability to apply for grants, and its ability to develop relationships across geographies and across sectors in a way that could be challenging for a single jurisdiction. With limited funding and as a voluntary member organization, the South Bay COG is an example of how a subregional agency can promote suburban sustainability.

SUBURBAN SUSTAINABILITY THROUGH SUBREGIONAL GOVERNANCE

Many suburban jurisdictions lack the capacity and resources to respond individually to state climate mandates and compete with larger cities for sustainability grants. In the absence of strong state or regional oversight, subregional agencies, such as councils of governments or county transportation authorities, can help localities efficiently respond to state and regional carrots and sticks for climate planning. Although they typically have limited power, subregional agencies can coordinate among cities to implement state and regional climate goals.

California's climate change framework, which lacks strong oversight of local actions but depends on local GHG emissions reductions from land use, benefits from subregional governance in the suburbs. In Southern California, subregional agencies representing county and subcounty geographies helped suburban jurisdictions engage in climate planning. These coordinated efforts helped suburbs avoid being sued by the state and compete collectively for state and regional grants. In the Inland Empire, subregional agencies negotiated multijurisdictional GHG reduction targets, measured local emissions, and developed tools for individual jurisdictions to incorporate emission reductions into their local general plans and climate action plans. In the South Bay in Los Angeles County, a subregional agency attracted state and regional grant dollars for local climate planning efforts and coordinated a private sector sustainable mobility pilot among multiple jurisdictions.

Long-term suburban sustainability will require addressing much larger questions of regional sustainability, such as revenue for transportation infrastructure and zoning reform to promote infill housing development near jobs, transit, and good schools. However, for large regions, these efforts will likely require building on existing forms of governance at the middle scale of subregional agencies. Looking beyond a singular focus on the metro scale, where big cities dominate the conversation, or on individual suburban jurisdictions with limited capacity, may reduce the unevenness, and increase the effectiveness, of the implementation of state climate goals.

ACKNOWLEDGMENTS

The author thanks Karen Chapple, Elizabeth Deakin, Margaret Weir, and Karen Christensen for their comments on this research. Thanks go to the U.S. Department of Transportation, the Bay Area Women's Transportation Seminar, and the Graduate Division and the Department of City and Regional Planning at the University of California, Berkeley for funding this research. All errors and opinions are solely the author's. The views expressed in this work do not necessarily represent those of the Federal Reserve Bank of San Francisco or of the Federal Reserve System.

NOTES

1. This case study is based on twenty semistructured interviews conducted by the author from 2014 to 2015 in person and by phone averaging an hour in length. Interviewees were selected because of their role at an organization involved in local climate planning in Southern California, recommendations from early interviewees (snowball sample), or because they agreed to be interviewed after taking a survey by the author on this topic.

2. EIRs are commonly called "environmental impact statements" (EISs) in other states.

3. General plans are commonly called "comprehensive plans" in other states.

4. Metropolitan regions in the U.S. with more than 50,000 people receiving federal transportation funding have an MPO to create and update a regional transportation plan (RTP). SCSs are a new chapter of the RTP in California that integrate state housing targets with transportation spending. For more detail on this, see Mattiuzzi 2016.

5. WRCOG administers the TUMF, local jurisdictions implement projects with the local return funds, and the Riverside County Transportation Commission (RCTC) programs and implements countywide projects.

6. Six jurisdictions prepared their own climate action plans; WRCOG assisted the other twelve.

REFERENCES CITED

Boarnet, Marlon G., Kenneth Joh, Walter Siembab, William Fulton, and Mai Thi Nguyen. 2011. "Retrofitting the Suburbs to Increase Walking: Evidence from a Land-Use-Travel Study." *Urban Studies* 48(1): 129–159.

Car2go. 2015. "Press Release: Car2go to Temporarily Suspend Service in the South Bay." Last accessed June 21, 2017. https://www.car2go.com/media/data/usa/microsite-press/files/car2go_la_south_bay_temp_suspension_press_release_-_final.pdf.

CEQA. n.d. California Environmental Quality Act Guidelines § 15152.

Chapple, Karen. 2015. *Planning Sustainable Cities and Regions: Towards More Equitable Development.* Routledge Equity, Justice and the Sustainable City Series. Oxfordshire, England; New York: Routledge.

Dryden, Carley. 2015. "Car2go Car-Sharing Service to Suspend South Bay Operations." *Daily Breeze*, May 4. Last accessed June 21, 2017. http://www.dailybreeze.com/business/20150504/car2go-car-sharing-service-to-suspend-south-bay-operations.

Dunham-Jones, Ellen, and June Williamson. 2011. *Retrofitting Suburbia, Updated Edition: Urban Design Solutions for Redesigning Suburbs.* Hoboken, NJ: John Wiley & Sons.

Dutton, Bob. 2007. *CEQA: Greenhouse Gas Emissions.* Senate Bill 97, Chapter 185, Statutes of 2007. Public Resources Code § 21083.05.

Elkind, Ethan. 2014. *Railtown: The Fight for the Los Angeles Metro Rail and the Future of the City.* Berkeley and Los Angeles: University of California Press.

Estiri, Hossein. 2016. "Differences in Residential Energy Use Between US City and Suburban Households." *Regional Studies* 50(11): 1919–1930.

Farren, Julie. 2017. "Beaumont City Council Approves Lawsuit Settlement Agreement with WRCOG." *Banning Record Gazette*, April 7. Last accessed February 14, 2020. https://www.recordgazette.net/news/local/beaumont-city-council-approves-lawsuit-settlement-agreement-with-wrcog/article_a1fc9172-1bea-11e7-bcdc-93aa1a0fde43.html .

Hanlon, Bernadette. 2009. *Once the American Dream: Inner-Ring Suburbs of the Metropolitan United States.* Philadelphia: Temple University Press.

Kolko, Jed. "Americans' Shift to the Suburbs Sped Up Last Year." 2017. *FiveThirtyEight.* March 23. Last accessed June 21, 2017. https://fivethirtyeight.com/features/americans-shift-to-the-suburbs-sped-up-last-year/.

LAO, California Legislative Analyst's Office. 2015. "California's High Housing Costs: Causes and Consequences." Last accessed June 21, 2017. http://www.lao.ca.gov/reports/2015/finance/housing-costs/housing-costs.pdf.

Mattiuzzi, Elizabeth. 2016. "Local Capacity for Implementing a State Climate Planning Mandate: The Politics of Cooperation and Regional Governance in California." PhD dissertation, University of California, Berkeley. http://www.planningsustainableregions.org/sites/default/files/2016_mattiuzzi_dissertation.pdf.

Pavley, Fran, and Fabian Nunez. 2006. *The California Global Warming Solutions Act.* Assembly Bill 32, Chapter 488, Statutes of 2006. Health and Safety Code § 38500-38598.

Rapino, Melanie A., and Alison K. Fields. 2013. "Time and Distance in Defining the Long Commute Using the American Community Survey." U.S. Census Bureau. Working paper 2013-03. Last accessed February 14, 2020. http://www.census.gov/hhes/commuting/files/2012/Paper-Poster_Megacommuting%20in%20the%20US.pdf.

Settlement Agreement. 2007. Settlement Agreement in the People of the State of California, ex. rel. Attorney General Edmund G. Brown Jr., versus County of San Bernardino, San Bernardino County Board of Supervisors. Last accessed February 14, 2020. https://oag.ca.gov/system/files/attachments/press_releases/2007-08-21_San_Bernardino_settlement_agreement.pdf.

Shaheen, Susan, Lauren Cano, and Madonna Camel. 2016. "Exploring Electric Vehicle Carsharing as a Mobility Option for Older Adults: A Case Study of a Senior Adult Community in the San Francisco Bay Area." *International Journal of Sustainable Transportation* 10(5): 406–417.

Siembab, Walter, Marlon Boarnet, Mohja Rhoads, Sabrina Sander, Frank Hotchkiss, Jacki Bacharach, Bill Fulton, and Amitabh Barthakur. 2009. "Sustainable South Bay: An Integrated Land Use and Transportation Strategy." Last accessed June 21, 2017. http://www.southbaycities.org/sites/default/files/documents/Sustainable%20South%20Bay%20Strategy.09.08.09_0.pdf.

Steinberg, Darrell. 2008. *Sustainable Communities and Climate Protection Act.* Senate Bill 375, Chapter 728, Statutes of 2008. Health and Safety Code § 65080, 65400, 65583, 65584, 65587, 65588, 14522, 21061, and 21159.

Wachs, Martin. 2003. "Local Option Transportation Taxes: Devolution as Revolution." *ACCESS Magazine* 1(22): 9–15. http://escholarship.org/uc/item/2d38m621.

Wikstrom, Nelson. 1977. *Councils of Governments: A Study of Political Incrementalism.* Chicago: Burnham.

WRCOG, Western Riverside Council of Governments. 2014. "Western Riverside Council of Governments Subregional Climate Action Plan." Last accessed February 14, 2020. http://www.wrcog.cog.ca.us/DocumentCenter/View/188/Subregional-Climate-Action-Plan-CAP-PDF?bidId=.

12

The Application of Land Use Regression and the National Land Cover Dataset in Modeling of Ozone Mixing Ratios in Baton Rouge, Louisiana

Chronic exposure to low concentrations of ambient air pollution is an economic and public health concern. Generally, concentrations of air pollutants are denser near and within urban centers: industry and traffic congestion load the atmosphere with anthropogenic emissions, while the physical structure of an urban center (e.g., street canyons) may discourage natural dispersion of emissions or create microclimates (e.g., urban heat island) that foster pollution. Suburban and rural regions surrounding urban centers are exposed to detrimental effects as pollution originating in urban centers moves downwind and augments localized pollution. In the case of ozone (O_3), the augmentation of precursory pollutants may act to prolong or exacerbate O_3 concentrations at low but still harmful levels in suburban and rural areas. As human geography evolves with population shifts, planning with a focus on mitigating pollution exposure becomes ever more pertinent to protect property, the ecosystem, and public health. High-resolution spatial predictions are critical to linking air pollution exposure to health outcomes, to establish and evaluate air quality policies, and to assess the impacts of land use and urban metabolism on air quality sustainability.

The purpose of this study is to downscale spatial predictions of complex, nonlinear air pollutants using easily accessible datasets thereby producing more informative maps more rapidly and with fewer financial constraints than by alternative means. Baton Rouge, Louisiana, is selected because the area has had a network of air quality monitors operating for many years. Importantly, observations from these monitors and documentation is available to the public through state and federal environmental agency websites.

More importantly, at least ten of the monitors have been collecting hourly observations since the year 2000. A 30 m resolution grid of classified land cover is used to downscale the spatial predictions of the O_3. The land cover data collection and classification are overseen by a consortium of governmental agencies and covers the entire contiguous United States. This data is available to the public.

In the effort to map air pollution and pollution exposure, land use regression (LUR) presents an improvement upon spatial interpolation. Spatial interpolation can achieve the finest spatial resolution where there is a dense, evenly distributed network of sampling points. Because suburban and rural rarely experience pollution levels great enough to exceed federal thresholds, and thus not require long-term monitoring, these areas lack a sufficient density of stations for spatial interpolation to achieve a high degree of spatial resolution.

LUR is based on the principle that the dependent variable at any location depends on the characteristics of the surrounding environment, particularly those that influence emission intensity and dispersion efficiency. LUR models have been applied to air quality studies to predict pollutant mixing ratios at locations remote from air quality sampling stations in the Western Europe (Briggs et al. 1997; Briggs et al. 2000; Beelen et al. 2007; Morgenstern et al. 2007; Rosenlund et al. 2008; Gulliver et al. 2011; Aguilera et al. 2013), Canada (Gilbert et al. 2005; Kanaroglou et al. 2005; Sahsuvaroglu et al. 2006; Henderson et al. 2007; Jerrett et al. 2007) and large U.S. metropolitan areas such as New York City (Ross et al. 2007) and Los Angeles (Moore et al. 2007). These studies focus on NO_2 and particulate matter pollution. O_3, a pollutant characteristic of photochemical smog, is considered by the EPA as a criteria pollutant, along with nitrogen oxides (NO_x), particulate matter, carbon monoxide, lead, and sulfur dioxide. Criteria pollutants are identified for their ubiquity and potential to cause harm.

Clinical studies and animal autopsies have confirmed that O_3 exposure induces adverse structural, functional, and biochemical alterations to biological tissues. The uptake of O_3 via stomatal gas exchange damages crops and forests (Feng et al. 2014) reducing agricultural yields (Krupa and Manning 1988; Fuhrer, Skärby, and Ashmore 1997; Avnery, Mauzerall, and Horowitz 2011). Inhalation of O_3 by humans can produce immediate respiratory problems, exacerbate asthma attacks, and increase the risk of respiratory infections and pulmonary inflammation (Beckett 1991). O_3 exposure damages the ocular surface (Lee et al. 2013) and increases the risk of a perforated appendix (Kaplan et al. 2013). O_3-induced illnesses decrease worker productivity (Zivin and Neidell 2013) and increase school absenteeism (Romieu et al. 1992; Gilliland et al. 2001; Hall, Brajer, and Lurmann 2003; Currie et al. 2009). In a review of O_3 research published between 2006 and 2012, the United States Environmental Protection Agency (2013) concluded that O_3 exposure is likely to cause cardiovascular harm, damage to the central nervous system, reproductive and developmental harm, and early death.

As of 2010, approximately 732,607 people lived in the Baton Rouge nonattainment zone (BRNZ), a five-parish area that struggles to meet federally established National Ambient Air Quality Standards (NAAQS) for O_3 pollution (United States Census Bureau 2010) exposing upwards of 16 percent of the state population to detrimental effects. In the early 1990s, the Baton Rouge area was classified as having a "serious" O_3 problem following the 1990 Clean Air Act criteria. Based on air quality data obtained by ten monitoring stations from 2011 to 2013, the USEPA determined the Baton Rouge area to be in attainment of the 2008 eight-hour O_3 NAAQS in April 2014. However, a stricter eight-hour O_3 NAAQS of 70 ppb was adopted in 2015, threatening Baton Rouge's O_3 attainment status.

Contributing to the persistent O_3 problem in the Baton Rouge area is a heterogeneous landscape of precursor emissions. Woody wetlands engulf a petrochemical corridor that follows the Mississippi River and interspersed pastures and forests bound the northeastern edge of an urban core. Nonlinear photochemistry produces O_3 from two classes of precursor emissions: volatile organic compounds (VOCs) and nitrogen oxides (NO_x, a collective term for nitrogen oxide [NO] and nitrogen dioxide [NO_2]). In 2011, 78.98 percent of NO_x emissions in the United States came from fossil fuel combustion for transportation, electricity generation, and industrial processes. While fossil fuel combustion releases VOCs, VOCs predominately originate from biogenic sources, in particular evergreen forests and citrus groves (Guenther 1997; Isebrands et al. 1999; Staudt and Kesselmeier 1999; Wagner and Kuttler 2014). According to the National Emissions Inventory (2011), biogenic sources accounted for 68.65 percent of all VOC emissions in the United States.

The complex chemistry of O_3 production and variegated land cover potentially exacerbate pollution exposure in a region larger than the urban center. Net O_3 production depends on the ratio of NO_x and VOCs and is suppressed when either is present in large enough excess relative to the other (Sillman 1999). Urban centers typically exhibit VOC sensitivity, where NO_x-related O_3 production is at a maximum and a lack of VOCs limits further production. By contrast, NO_x sensitivity is characteristic of rural areas. In such situations, VOC-related O_3 production is stalled because of a lack of NO_x. Chemistry inside an air mass evolves as it moves over areas of different emissions. For example, O_3 production that has ceased in a NO_x-saturated air mass overlying an urban center may resume as the air mass drifts into less industrial, dense, more rural areas where VOCs are more plentiful. Generally, VOC emissions control the initial buildup of O_3 while NO_x emissions from within an urban area determine the total amount of ozone that is formed after the air moves downwind (Sillman 1999).

Predictive variables in LUR studies most often fall into five categories: land use, physical geography, meteorology, roads and traffic, and population. LUR uses the large geographic extent and/or sampling density of these independent variables to downscale the spatial predictions of the dependent variable, which is O_3 in this case. LUR results may stand alone or may be combined with spatial interpolation in a hybrid regression-interpolation technique. Independent variables attempt to capture the range of temporal and spatial scales of mixing ratios and precursor emission sources and dispersion. In this case study, the objective is a simple model not affected by data availability so it may be replicated many times over.

Given the long-running problem with O_3 pollution and the varied landscape the Baton Rouge area is chosen for this case study where the aim is to develop a LUR.

Data and Methods

Study Area

The study area boundaries match those of the Louisiana Department of Environmental Quality (LDEQ) Capital Region. Ascension, East Baton Rouge (EBR), Iberville, Livingston, and West Baton Rouge (WBR) parishes, which comprise the five-parish Baton Rouge nonattainment zone (BRNZ), are part of the LDEQ Capital Region as are ten air quality monitors (Figure 12.1).

Located within the LDEQ region are the cities of Port Allen and Baton Rouge. To the southeast of the study are is New Orleans. These three cities are connected west to east by Interstate 10 and suburban development.

Land Cover as the Independent Variable

The Multi-Resolution Land Characteristics Consortium (MRLC), a collaboration of federal agencies, maintains the National Land Cover Database (NLCD) (Homer et al. 2007; Fry et al. 2011; Homer et al. 2015). NLCD

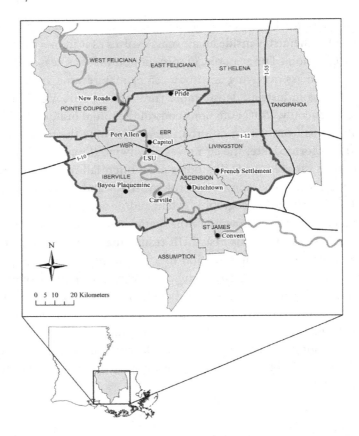

Figure 12.1. Study area including the five parish nonattainment zone (*heavy gray line*) and ten air quality monitoring sites.

products provide consistent, nationwide land cover information at a 30 m resolution for scientific, economic, and governmental applications. Mosaic Landsat imagery is classified via a decision-tree algorithm. Each 30 m grid is assigned one of sixteen possible designations (Table 12.1). NLCDs are available for 2001 (NLCD 2001), 2006 (NLCD 2006), and 2011 (NLCD 2011). Fifteen of the sixteen land cover types appear within the study area.

To investigate the potential explanatory ability of different land covers, univariate regression analysis was conducted whereby a proportion for a land cover type within a buffer radius was regressed against O_3. Around each of the 10 air quality monitors buffers of radii 100, 200, 300, 400, 500, 1,000, 1,500, 2,000, 2,500, 3,000, 3,500, 4,000, 4,500, 5,000, 7,500, and 10,000 m were computed. Within each buffer, the proportion of each of the land cover classes was computed with the exclusion of perennial ice/snow which

Table 12.1. Land cover classes and abbreviated codes

Class	Code	Class	Code
Open water	WA	Evergreen forest	EV
Perennial ice/snow	—	Mixed forest	MI
Developed, open space	OS	Shrub/scrub	SH
Developed, low intensity	LO	Grassland/herbaceous	HE
Developed, medium intensity	ME	Hay/pasture	HA
Developed, high intensity	HI	Cultivated crops	CR
Barren land (rock/sand/clay)	BA	Woody wetlands	WW
Deciduous forest	DE	Emergent herbaceous wetlands	EH

did not appear in the study area. This was repeated for NLCD 2001, NLCD 2006, and NLCD 2011 thereby increasing the number of trials from 10 to 30 for more statistically robust results.

O_3 as the Dependent Variable

Daily maximum 8-h mean O_3 mixing ratios recorded by 10 air quality monitoring stations operating within or near the BRNZ were collected from the USEPA (Table 12.2). Data completeness at the 10 sites averages 97.5 percent. Algorithms for counting a day as having complete data mirror those used by USEPA. The specific algorithm is described as follows.

The daily maximum eight-hour O_3 mixing ratio for a given day is the highest of the 24 possible eight-hour mean mixing ratios computed for that calendar day. Running eight-hour averages are computed from the hourly

Table 12.2. The ten air quality monitoring stations in the study area with latitude and longitude

Name	Latitude	Longitude
Bayou Plaquemine	30.220556	-91.316110
Capitol	30.461980	-91.179220
Carville	30.206985	-91.129948
Convent	29.994444	-90.820000
Dutchtown	30.233889	-90.968333
French Settlement	30.312500	-90.812500
LSU	30.419763	-91.181996
New Roads	30.681736	-91.366172
Port Allen	30.500643	-91.213556
Pride	30.700921	-91.056135

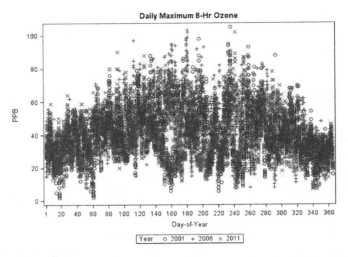

Figure 12.2. Daily maximum eight-hour ozone observations for 2001, 2006, and 2011 plotted by day of year.

O_3 mixing ratio data and the result is stored in the first hour of the eight-hour period. An eight-hour average is considered valid if at least six of eight hourly averages for the eight-hour period are available. An O_3 monitoring day is counted as a valid day if valid eight-hour averages are available for at least 18 of 24 possible hours in the day. If fewer than 18 of the eight-hour averages are available, a day is counted as a valid day if the daily maximum eight-hour average mixing ratio for that day exceeds the ambient standard.

Daily maximum eight-hour O_3 mixing ratios in the study area display a seasonal trend with maxima during the northern hemisphere summer months exceeding those reached in early spring and winter (Figure 12.2). Interestingly, the data for 2001 dips around Julian day 160, possibly due to the development of Tropical Storm Allison in the northwestern Gulf of Mexico on June 5, 2001. Here, the Julian day is the count of the day from the beginning of the year with January 1 being Julian day 1. O_3 mixing ratios within a few hundred km of an intensifying storm have been observed to decrease (Zou and Wu 2005). The seasonal trend as well as this dip is resolved by detrending the data (Figure 12.3).

Synoptic-scale meteorology may confound local-scale variability in O_3 mixing ratios. Abraham and Comrie (2004) removed and retained the regional trend from O_3 observations in Tucson to overcome limited meteorological data. Since official meteorological data for the study area is recorded only at the airport which is displaced from any air quality monitor, this

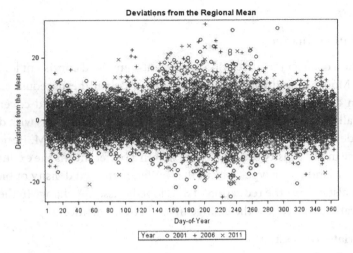

Figure 12.3. Daily maximum eight-hour ozone observations for 2001, 2006, and 2011 with the daily regional mean removed.

study follows the example of Abraham and Comrie (2004) in using deviations-from-the-regional-mean (devRM) rather than direct observations. This process also emphasizes site-specific variability, strengthening the correlations of land cover variables.

After removal of the regional mean, the number of significantly correlated predictors and the overall strength of correlations increases for daily maximums, monthly average maximums, and triennial monthly average maximums. For daily O_3 observations, the number of significantly correlated predictors increases from 205 to 227 and the mean absolute correlation coefficient improves from 0.055 to 0.146. For monthly averages, the number of significantly correlated predictors increases from 111 to 204 and the mean absolute correlation coefficient improves from 0.133 to 0.286. For a triennial monthly average, the number of significantly correlated predictors increases from 110 to 207 and the mean absolute correlation coefficient improves from 0.132 to 0.306.

Multiple Regression Model Development

Land cover variables were tested for significance at 95 percent confidence ($p<0.05$). The buffer size with the highest adjusted R^2 (Adj R^2) for each land cover was identified and entered into the model in a stepwise selection. To enter the model, a predictor must have improved the R2 value by at least one percent.

RESULTS

Correlation Analysis

Direction of effect for correlation coefficients remained consistent for Model B and Model C, and with the exception of grassland/herbaceous (HE), direction did not change among buffer radii within each land cover class. Generally, biogenic land cover classes had positive correlations with devRM while developed classes had negative correlations with devRM. Because an increase in the density of a land cover within a buffer means the exclusion of the other land cover types within that buffer, increased density of biogenic classes also means the reduction of developed classes. A change in the NO_x-VOC ratio could follow the change in land cover partitioning.

Univariate Regression

A series of univariate regressions reduced the field of candidate predictors to a single buffer for each land cover. An "all-in" approach without filtering of candidate predictor variables has the potential for greater explained variance but runs the risk of overfitting the model and of using redundant information when multiple buffers for a given land cover enter the model.

Multiple Regression

Results of the univariate regressions informed a stepwise selection. Table 12.3 shows the results of the stepwise selection.

Predictors in the stepwise selection model, particularly LO, ME, and HI, were correlated as seen in Table 12.4. Variance inflation factors (VIFs) were examined to control for the danger of having too much correlation among predictors, or multicollinearity. A large VIF is often used as a sign of multicollinearity, which can limit the conclusions that can be drawn from the regression coefficients about the contribution of each covariate (Zainodin and Yap 2013).

To reduce model redundancy, any predictor with a VIF greater than five was removed, resulting in the elimination of the developed–medium (ME) and shrub/scrub (SH) predictors from Model A. ME was removed from Model B and developed–low (LO) was removed from Model C. Additionally, deciduous forest (DE) was removed from Model A due to nonsignificance. The final models and parameter estimates are in Table 12.5.

Next, the models were evaluated for normality, and for fit using the Adj R^2, the root mean square error (RMSE), and plots of the residuals. Model C achieved the greatest Adj R^2 value at 0.4204, with a RMSE of 1.95618. Model

Table 12.3. Initial stepwise selection results

Model A:	-8.2616-61.0504 * he01500 + 74.6579 * ev04500 + 11.7719 * ha00100-42.9699 sh04500-22.7764 * hi00100 + 66.6467 * me07500 + 6.5643 * cr00200 + 14.1712 * ww00400 + 100.6700 * de02500
Model B:	-2.0950 + 46.3469 * he1500 17.6080 * ev4500 + 2.8860 * ha0100-12.8654 * hi100 + 17.2481 * me7500
Model C:	-0.9648 + 3.0209 * ha0100 + 17.9599 * ev10000-18.6309 * hi0100-14.2063 * lo04500 + 41.3818 * me07500

Table 12.4. Developed land cover types (HI, ME, and LO) are highly correlated to each other

Pearson Correlation Coefficients, N=360	HA00100	EV10000	HI00100	ME07500
HA00100	1.0000	0.4762	-0.4964	-0.7282
		<.0001	<.0001	<.0001
EV10000	0.4762	1.0000	-0.2576	-0.3921
	<.0001		<.0001	<.0001
HI00100	-0.4964	-0.25764	1.0000	0.7879
	<.0001	<.0001		<.0001
ME07500	-0.7282	-0.3921	0.7879	1
	<.0001	<.0001	<.0001	
LO04500	-0.7991	-0.4824	0.6444	0.9313
	<.0001	<.0001	<.0001	<.0001

Table 12.5. Final multiple regression models

Model A:	-0.4433 + 0.4543 * he01500 + 0.189 * ev04500 + 0.0076 * ha00100-0.0736 * hi00100—cr00200 * 0.0179-0.0197 * ww00400
Model B:	-1.0184 + 0.3861 * he01500 + 0.1782 * ev04500 + 0.0145 * ha00100-0.0631 * hi 00100
Model C:	-1.9325 + 0.0387 * ha00100 + 0.2106 * ev10000-0.1576 * hi00100 + 0.2121 * me07500

B had an Adj R^2 of 0.2905 and a RMSE of 2.35759. For Model A the Adj R^2 is 0.0956, and a RMSE of 4.90205. Model A did not pass a modified Kolmogorov-Smirnov normality test conducted in SAS 9.4 (Figure 12.4; Table 12.6). For samples with fewer than 2,000 observations, SAS 9.4 outputs the Shapiro-Wilk statistic. Models B (Figure 12.5; Table 12.7) and C (Figure 12.6; Table 12.8) passed the Shapiro-Wilk test for normality.

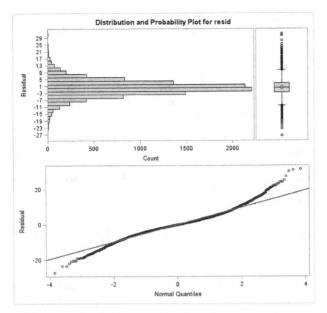

Figure 12.4. Model A residual plots. Notice the skewing on the tails.

Table 12.6. Model A normality test results

Tests for Normality

Test		Statistic	p Value
Kolmogorov-Smirnov	D	0.05287	Pr >D

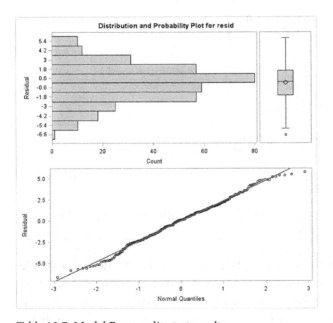

Figure 12.5. Model B residual plots.

Table 12.7. Model B normality test results

Tests for Normality

Test		Statistic	p Value
Shapiro-Wilk	W	0.993877	Pr >W

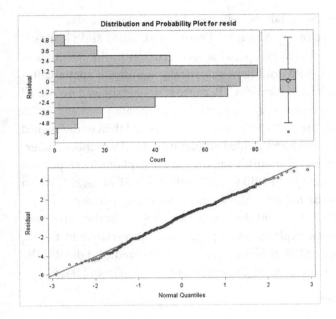

Figure 12.6. Model C residual plots.

Table 12.8. Model C normality test results

Tests for Normality

Test		Statistic	p Value
Shapiro-Wilk	W	0.993094	Pr >W

Given the poor performance and lack of normality in the residuals, Model A was excluded from further analysis. The predictive abilities of the model were also evaluated using the prediction sum of squares (PRESS) statistics, equivalent to "leave one out" cross validation (LOOCV), and RM-SEs (Table 12.9). Both models exhibit variance in the tails. Because Model C achieved both a lower RMSE and PRESS than Model B, Model C was used to produce a surface.

Table 12.9. Performance and error statistics for Model B and Model C

	Adj R²	PRESS	RMSE
Model B	.2905	2036.2998	2.3596
Model C	.4204	1396.7001	1.9562

Note: Model A was excluded due to poor normality.

Spatial autocorrelation analysis indicates that Model C was able to capture the spatial nature of O_3. Moran's I (Wang et al. 2015) and Geary's C (Shaker et al. 2010) revealed no spatial autocorrelation in the residuals of Model C. The mean devRM did have statistically significant spatial autocorrelation by Moran's I and Geary's C. Because residuals showed no spatial autocorrelation, a regression-only approach is appropriate.

Model C applied to the 2010–2012 monthly means and then to the pooled data across all years considered yielded similar explained variances and error statistics. The three-year monthly regional mean explains 83.68 percent of variance in the O_3 data from 2010 to 2012 with an RMSE of 2.4293. When summed with the predicted devRM, explained variance increases to 91.25 percent and the RMSE is 1.7790. For all periods together, the three-year monthly regional mean explains 88.65 percent of the variance in the O_3 data with an RMSE of 2.5729. When summed with the predicted devRM the explained variance increases to 93.50 percent with an RMSE of 1.9479.

Surface Computation

Model C applied to the study area produces a 30 m resolution grid of estimated devRM. Results for NLCD 2011 range from -14.8322 to 9.48497 ppb (Figure 12.7). To this surface may be added the regional mean computed for any day within the 2010–2012 period. Seasonality is captured in the regional mean which was removed and retained. For comparison, a surface generated with inverse distance weighting (IDW) (Figure 12.8), a commonly applied interpolation technique whereby estimates are a spatially weighted average, lacks the spatial character of the LUR surface.

Discussion and Conclusions

Correlation Analysis

Under the assumption that biogenic classes are sources of VOCs and that developed classes are sources of NO_x, the direction of effect observed in the correlation analysis could be explained by the VOC to NO_x ratio and would fit the observation downtown-to-downwind evolution of an air mass from VOC sensitivity to NO_x sensitivity. The strength of correlations among the developed classes revealed by VIFs and correlation analysis, especially between LO and ME, warrants further examination.

The MRLC (MRLC 2016) defines these classes as "areas with a mixture of constructed material and vegetation . . . These areas most commonly

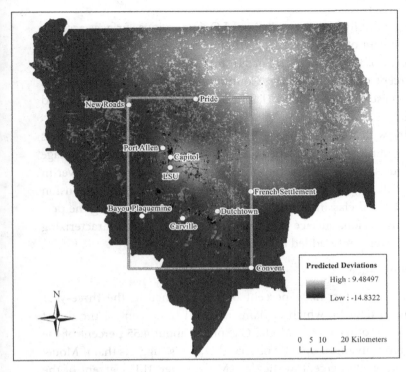

Figure 12.7. Prediction surface generated with Model C.

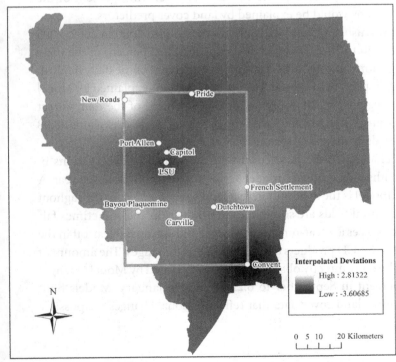

Figure 12.8. Surface generated with IDW.

include single-family housing units." For LO, impervious surfaces account for 20–49 percent of total cover. For ME, impervious surfaces account for 50–79 percent of the total cover. For HI, impervious surfaces account for 80–100 percent of the total cover. Continuous data would eliminate the conflict of these developed land classifications that are based on binned percentages.

Beginning with NLCD 2001, the MRLC began producing two continuous datasets in addition to the classified dataset: one measuring the percentage of impervious surfaces and another for the percentage of canopy cover in each 30 m cell. So, developed land indicators will have more dimension (from four binned classifications to 1 percent intervals) by using the percentage of impervious surface raster, and the separate layer characterizing the canopy cover can be added to the analysis.

Model Performance

Model C accounts for 42.04 percent of the variance in the three-year monthly mean devRM, which explains about 11.12 percent of the direct observations. In other words, Model C explains about 4.55 percent of the three-year monthly mean directly observed O_3. This suggests that if Model C could capture 100 percent of the devRM data then 11.12 percent of the direct observations would be explained by land cover predictors.

Model C was also applied to daily observations from 2000 to 2002, 2005 to 2007, and 2010 to 2012. Daily devRM explains 12.22 percent of the daily direct observations, and Model C captures 9.93 percent of the daily devRM, resulting in 1.20 percent of the daily direct observations explained by Model C. The daily regional mean accounts for 87.78 percent of the daily direct observations, and when summed with the predicted devRM, explains 88.98 percent of the daily direct observations.

Seasonal changes in the land cover and the spread of observations in summer months may influence model performance throughout the year. A pitfall of the model is the assumption that land cover is constant throughout a year. Agricultural fields are sometimes in production and sometimes fallow; deciduous trees are leaf-on in the summer and spring but leaf-off in the fall and winter. NLCD products do not reflect these changes. The amount of variance in the daily directly observed O_3 data explained by Model C ranges from 0.27 percent in September to 6.21 percent in January. Models built by month and/or land cover data that reflect seasonal changes are possible solutions.

Predicted Surface

Within the study area, there is an overall spatial trend of negative devRM in the southwest to positive devRM in the northeast, with the greatest negative devRM predicted in urbanized areas. Urban centers typically struggle with exceedances more than suburban or rural areas yet in this study the urban areas are predicted to have departures below the regional average (i.e., lower O_3 mixing ratios). The length of the averaging period could explain some of this incongruence between what might be expected based on historic exceedances and the predicted surface. O_3 events occur on the scale of hours and days, not months, and NO_x-loading events such as rush hour traffic may induce spikes of O_3 production. Such small-scale variation gets "washed out" when averaged over a month, but frequent mildly elevated O_3 mixing ratios would appear as a positive deviation over a long averaging period. Additionally, a large NO_x to VOC ratio, characteristic of urban areas, can induce the NO-titration whereby O_3 is destroyed (Sillman 1999).

The LSU, Capitol, and Port Allen sites exhibit a large displacement of the mean from the median relative to the other sites and extreme observations above the upper fences in box plots, but these sites frequently have the lowest means. In the case of positive skew, extreme observations can pull the mean in the positive direction despite a majority of observations occurring below the mean value. In a comparison of an urban traffic site, a suburban site, and a rural site, Im et al. (2013) observed greater fluctuations throughout the day at the urban site than at the rural site, but a greater monthly mean O_3 mixing ratios at the rural site.

Generally, the smaller statistical means at the LSU, Capitol, and Port Allen sites are maintained, and the IQR is smaller than other stations. This indicates that on most days the maxima monitored by these sites deviates little from the regional mean maximum. This indicates nothing about the severity of extreme observations at the sites considered since on exceedance days the regional mean may be large and only a slight deviation would result in an exceedance. Conversely, large deviations might occur on days where the regional mean is low and thus not exceeding the design value.

Future Research

The results of this study are best interpreted as the likelihood of chronic exposure to elevated levels of O_3, levels that do not necessarily exceed NAAQS. Negative deviations should not be considered a decrease in the O_3 in those areas, but that such areas, on average, experience daily O_3 maxima below that of the region. Predictions say nothing about the relative severity of daily mixing ratios at individual sites. The monthly results of this study could aid epidemiologists investigating health effects (asthma, birth related, etc.) of long-term exposure.

Future attempts to model O_3 in the Baton Rouge area should include variables with more temporal resolution. Synoptic-scale meteorology is captured in the regional mean, but local wind conditions may provide insight on the transport of precursors. Light winds in the Baton Rouge area (NOAA 2011) cause drift rather than dispersion. Winds blowing over NO_x-saturated urbanized areas may carry these O_3 precursors into rural regions rich in VOCs. Traffic patterns could characterize the time of expected NO_x loadings as well as the geography of emissions.

This study focused only on land cover and did not account for land use. Land cover may be able to capture some latent variables related to surface thermal properties, dispersion, and emissions. Land use may be better at characterizing the geography of precursor emissions. For example, a point source such as a petrochemical plant surrounded by agricultural land or forests would not be captured by land cover. Due to the resolution of the raster, roadways may not be captured, and for those that are, there is no indication of how much the roadways are trafficked. An interstate corridor is more intensely used than a rural road. Many studies use either roadway classifications or vehicle miles traveled for this reason.

Using auxiliary information in interpolating phenomena helps refine the spatial resolution. Localized spatial character was emphasized by removing the regional mean, or shared variance, from the point-based observations. Land cover data did not perform well in predicting the daily maximum O_3 but performed moderately well for longer averaging periods. For monthly mean maximums, evergreen and developed classes were important predictors. Evergreen has a positive relationship while developed classes have negative relationships with the devRM. Given the strength of correlations and the importance of these variables in regression, NCLD products containing the percentage of impervious surface and percentage of canopy cover per pixel should be evaluated as potential predictors to represent the interspersion of land cover types in suburban areas.

More robust verification of results is desirable, as is the evaluation of the model performance at different sites and by season. The study could be expanded with data from similar climate regions, and the stability of the model tested with data from those regions. Independent sampling at locations between monitoring sites would improve validation statistics. While meteorology was not considered in this study, meteorological conditions at the time that daily O_3 maxima were reached could be evaluated and potentially incorporated as model parameters.

Despite the limitations of this study, it serves as a useful baseline in protecting life and property from the hazards of chronic exposure to O_3. The use of a nationwide land cover dataset means that the model may be used in places where long periods of O_3 observations are not available. Because the sources of variation are decomposed, the model is adaptable and expandable. The highly resolved surface achievable allowed for more precise research to be conducted in the realms of public health and suburban sustainability.

REFERENCES CITED

Abraham, J. S., and A. C. Comrie. 2004. "Real-Time Ozone Mapping Using a Regression-Interpolation Hybrid Approach, Applied to Tucson, Arizona." *Journal of the Air and Waste Management Association* 54: 914–925.

Aguilera, I., M. Pedersen, R. Garcia-Esteban, F. Ballester, M. Basterrechea, A. Esplugues, A. Fernandez-Somoano, A. Lertxundi, A. Tardon, and J. Sunyer. 2013. "Early-Life Exposure to Outdoor Air Pollution and Respiratory Health, Ear Infections, and Eczema in Infants from the INMA Study." *Environmental Health Perspectives* 121: 387–392.

Avnery, S., D. L. Mauzerall, J. Liu, and L. W. Horowitz. 2011. "Global Crop Yield Reductions Due to Surface Ozone Exposure: 1. Year 2000 Crop Production Losses and Economic Damage." *Atmospheric Environment* 45: 2284–2296.

Beckett, W. S. 1991. "Ozone, Air Pollution, and Respiratory Health." *Yale Journal of Biology and Medicine* 64: 167–175.

Beelen, R., G. Hoek, P. Fischer, P. A. van den Brandt, and B. Brunekreef. 2007. "Estimated Long-Term Outdoor Air Pollution Concentrations in a Cohort Study." *Atmospheric Environment* 41: 1343–1358.

Briggs, D. J., S. Collins, P. Elliott, P. Fischer, S. Kingham, E. Lebret, K. Pryl, H. Van Reeuwijk, K. Smallbone, and A. Van Der Veen. 1997. "Mapping Urban Air Pollution Using GIS: A Regression-Based Approach." *International Journal of Geographical Information Science* 11: 699–718.

Briggs, D. J., C. de Hoogh, J. Gulliver, J. Wills, P. Elliott, S. Kingham, and K. Smallbone. 2000. "A Regression-Based Method for Mapping Traffic-Related Air Pollution: Application and Testing in Four Contrasting Urban Environments." *Science of the Total Environment* 253: 151–167.

Currie, J., E. A. Hanushek, E. M. Kahn, M. Neidell, and S. G. Rivkin. 2009. "Does Pollution Increase School Absences?" *Review of Economics and Statistics* 91: 682–694.

Feng, Z., J. Sun, W. Wan, E. Hu, and V. Calatayud. 2014. "Evidence of Widespread Ozone-Induced Visible Injury on Plants in Beijing, China." *Environmental Pollution* 193: 296–301.

Fry, J., G. Xian, S. Jin, J. Dewitz, C. Homer, L. Yang, C. Barnes, N. Herold, and J. Wickham. 2011. "Completion of the 2006 National Land Cover Database for the Conterminous United States." *Photogrammetric Engineering & Remote Sensing* 77: 858–864.

Fuhrer, J., L. Skärby, and M. R. Ashmore. 1997. "Critical Levels for Ozone Effects on Vegetation in Europe." *Environmental Pollution* 97: 91–106.

Gilbert, N. L., M. S. Goldberg, B. Beckerman, J. R. Brook, and M. Jerrett. 2005. "Assessing Spatial Variability of Ambient Nitrogen Dioxide in Montreal, Canada, with a Land-Use Regression Model." *Journal of the Air & Waste Management Association* 55: 1059–1063.

Gilliland, F. D., K. Berhane, E. B. Rappaport, D. C. Thomas, E. Avol, W. J. Gauderman, S. J. London, H. G. Margolis, R. McConnell, K. T. Islam, and J. M. Peters. 2001. "The Effects of Ambient Air Pollution on School Absenteeism Due to Respiratory Illnesses." *Epidemiology* 12: 43–54.

Guenther, A. 1997. "Seasonal and Spatial Variations in Natural Volatile Organic Compound Emissions." *Ecological Society of America* 7: 34–45.

Gulliver, J., K. de Hoogh, D. Fecht, D. Vienneau, and D. J. Briggs. 2011. "Comparative Assessment of GIS-Based Methods and Metrics for Estimating Long-Term Exposures to Air Pollution." *Atmospheric Environment* 45: 7072–7080.

Hall, J. V., V. Brajer, and F. W. Lurmann. 2003. "Economic Valuation of Ozone-Related School Absences in the South Coast Air Basin of California." *Contemporary Economic Policy* (21): 407–417.

Henderson, S. B., B. Beckerman, M. Jerrett, and M. Brauer. 2007. "Application of Land Use Regression to Estimate Long-Term Concentrations of Traffic-Related Nitrogen Oxides and Fine Particulate Matter." *Environmental Science & Technology* 41: 2422–2428.

Homer, C., J. Dewitz, J. Fry, M. Coan, N. Hossain, C. Larson, N. Herold, A. McKerrow, J. N. VanDriel, and J. Wickham. 2007. "Completion of the 2001 National Land Cover Database for the Conterminous United States." *Photogrammetric Engineering and Remote Sensing* 73: 337–341.

Homer, C. G., J. A. Dewitz, L. Yang, S. Jin, P. Danielson, G. Xian, J. Coulston, N. D. Herold, J. D. Wickham, and K. Megown. 2015. "Completion of the 2011 National Land Cover Database for the Conterminous United States-Representing a Decade of Land Cover Change Information." *Photogrammetric Engineering and Remote Sensing* 81: 345–354.

Im, U., S. Incecik, M. Guler, A. Tek, S. Topcu, Y. S. Unal, O. Yenigun, T. Kindap, M. T. Odman, and M. Tayanc. 2013. "Analysis of Surface Ozone and Nitrogen Oxides at Urban, Semi-Rural and Rural Sites in Istanbul, Turkey." *Science of the Total Environment* 443: 920–931.

Isebrands, J. G., A. B. Guenther, P. Harley, D. Helmig, L. Klinger, L. Vierling, P. Zimmerman, and C. Geron. 1999. "Volatile Organic Compound Emission Rates from Mixed Deciduous and Coniferous Forests in Northern Wisconsin, USA." *Atmospheric Environment* 33: 2527–2536.

Jerrett, M., M. A. Arain, P. Kanaroglou, B. Beckerman, D. Crouse, N. L. Gilbert, J. R. Brook, N. Finkelstein, and M. M. Finkelstein. 2007. "Modeling the Intraurban Variability of Ambient Traffic Pollution in Toronto, Canada." *Journal of Toxicology and Environmental Health, Part A* 70: 200–212.

Kanaroglou, P. S., M. Jerrett, J. Morrison, B. Beckerman, M. A. Arain, N. L. Gilbert, and J. R. Brook. 2005. "Establishing an Air Pollution Monitoring Network for Intra-Urban

Population Exposure Assessment: A Location-Allocation Approach." *Atmospheric Environment* 39: 2399–2409.

Kaplan, G. G., D. Tanyingoh, E. Dixon, M. Johnson, A. J. Wheeler, R. P. Myers, S. Bertazzon, V. Saini, K. Madsen, S. Ghosh, and P. J. Villeneuve. 2013. "Ambient Ozone Concentrations and the Risk of Perforated and Nonperforated Appendicitis: A Multicity Case-Crossover Study." *Environmental Health Perspectives* 121(8): 939–943.

Krupa, S. V., and W. J. Manning. 1988. "Atmospheric Ozone: Formation and Effects on Vegetation." *Environmental Pollution* 50: 101–137.

Lee, H., E. K. Kim, S. W. Kang, J. H. Kim, H. J. Hwang, and T.-i. Kim. 2013. "Effects of Ozone Exposure on the Ocular Surface." *Free Radical Biology and Medicine* 63: 78–89.

Moore, D. K., M. Jerrett, W. J. Mack, and N. Kunzli. 2007. "A Land Use Regression Model for Predicting Ambient Fine Particulate Matter Across Los Angeles, CA." *Journal of Environmental Monitoring* 9: 246–252.

Morgenstern, V., A. Zutavern, J. Cyrys, I. Brockow, U. Gehring, S. Koletzko, C. P. Bauer, D. Reinhardt, H. E. Wichmann, and J. Heinrich. 2007. "Respiratory Health and Individual Estimated Exposure to Traffic-Related Air Pollutants in a Cohort of Young Children." *Occupational and Environmental Medicine* 64: 8–16.

MRLC. 2016. *National Land Cover Database: Product Legend.* Last accessed May 20, 2016. https://www.mrlc.gov/data/legends/national-land-cover-database-2011-nlcd2011-legend.

National Emissions Inventory. 2011. Air Emissions Inventory. https://www.epa.gov/air-emissions-inventories. Last accessed April 2, 2016.

NOAA. 2011. *Local Climatological Data Annual Summary With Comparative Data.* Baton Rouge, LA.

Romieu, I., M. C. Lugo, S. R. Velasco, S. Sanchez, F. Meneses, and M. Hernandez. 1992. "Air Pollution and School Absenteeism Among Children in Mexico City." *American Journal of Epidemiology* 136: 1524–1531.

Rosenlund, M., F. Forastiere, M. Stafoggia, D. Porta, M. Perucci, A. Ranzi, F. Nussio, and C. A. Perucci. 2008. "Comparison of Regression Models with Land-Use and Emissions Data to Predict the Spatial Distribution of Traffic-Related Air Pollution in Rome." *Journal of Exposure Science & Environmental Epidemiology* 18: 192–199.

Ross, Z., M. Jerrett, K. Ito, B. Tempalski, and G. D. Thurston. 2007. "A Land Use Regression for Predicting Fine Particulate Matter Concentrations in the New York City Region." *Atmospheric Environment* 41: 2255–2269.

Sahsuvaroglu, T., M. A. Arain, P. Kanaroglou, N. Finkelstein, B. Newbold, M. Jerrett, B. Beckerman, J. R. Brook, M. Finkelstein, and N. L. Gilbert. 2006. "A Land Use Regression Model for Predicting Ambient Concentrations of Nitrogen Dioxide in Hamilton, Ontario, Canada." *Journal of the Air & Waste Management Association* 56: 1059–1069.

Shaker, R. R., A. I. Crăciun, and I. Grădinaru. 2010. "Relating Land Cover and Urban Patterns to Aquatic Ecological Integrity: A Spatial Analysis." *Geographia Technica* 9: 76–90.

Sillman, S. 1999. "The Relation Between Ozone, NO_x and Hydrocarbons in Urban and Polluted Rural Environments." *Atmospheric Environment* 33: 1821–1845.

Staudt, M., and J. Kesselmeier. 1999. "Biogenic Volatile Organic Compounds (VOC): An

Overview on Emission, Physiology and Ecology." *Journal of Atmospheric Chemistry* 33: 23–88.

United States Census Bureau. 2010. Last accessed April 2, 2016. https://www.census.gov/quickfacts/fact/table/US/PST045219.

United States Environmental Protection Agency. 2013. *Integrated Science Assessment of Ozone and Related Photochemical Oxidants*. Washington, D.C.: U.S. Environmental Protection Agency.

Wagner, P., and W. Kuttler. 2014. "Biogenic and Anthropogenic Isoprene in the Near-Surface Urban Atmosphere—a Case Study in Essen, Germany." *Science of the Total Environment* 475: 104–115.

Wang, W., Y. Ying, Q. Wu, H. Zhang, D. Ma, and W. Xiao. 2015. "A GIS-Based Spatial Correlation Analysis for Ambient Air Pollution and AECOPD Hospitalizations in Jinan, China." *Respiratory Medicine* 109: 372–378.

Zainodin, H. J., and S. J. Yap. 2013. "Overcoming Multicollinearity in Multiple Regression Using Correlation Coefficient." *AIP Conference Proceedings* 1557: 416–419.

Zivin, J., and M. Neidell. 2013. "Environment, Health, and Human Capital." *Journal of Economic Literature* 51: 689–730.

Zou, X., and Y. Wu. 2005. "On the Relationship Between Total Ozone Mapping Spectrometer (TOMS) Ozone and Hurricanes." *Journal of Geophysical Research-Atmospheres* 110: 1–15.

IV

SUSTAINABLE LAND AND WATER MANAGEMENT IN THE SUBURBS

13

Leading through Water

Defining Sustainability through Leadership, Experience, and Engagement in the Pittsburgh Metropolitan Region, Pennsylvania

MICHAEL H. FINEWOOD AND SEAN MCGREEVEY

Although not necessarily a new conceptual framework at the time, "sustainability" gained popularity in the 1980s as a way to rectify a fraught relationship between the human and nonhuman world. Preceding development strategies far too often emphasized economic goals to the detriment of communities and the landscapes they live in. Likewise, conservation efforts were frequently critiqued for privileging nonhuman species and habitats over human needs. Efforts to merge these potentially disparate frameworks converged with the Brundtland Commission, where participants sought to conceptualize sustainability as an explicit relationship between human well-being and environmental health (Carr and Finewood 2015).

Emerging from the now famous Brundtland report (World Commission on Environment and Development 1987), today sustainability is perhaps best characterized as meeting the needs of the present without compromising the needs of future generations. The definition of sustainability has no doubt evolved since the 1980s; scholars and practitioners have drawn on or explicitly integrated ideas about security (Allouche 2011), systems (Meadows 2008), resilience (Redman 2014), and more to make the term more applicable to specific contexts (Vercoe and Brinkmann 2010). Various metaphors or representations include three-legged stools or concentric circles that demonstrate the coequal importance of economies, societies, and the environment in contributing to a healthy whole. More recent iterations have reminded us that we should not only be focusing on generations in the future (intergenerational equity) while communities here and now (intragenerational equity) are experiencing inequity and injustice as it relates to environmental change (Heynen, Perkins, and Roy 2006). True sustainability, in this view, recognizes the importance of now *and* the future.

Although each new iteration of sustainability advances our conversation, particularly those that seek to correct previous gaps, critiques still remain. Most commonly, sustainability is panned for being too broad, vague, or nebulous. For example, when sustainability is articulated without defining goals or assets, questions often arise as to what is actually being sustained and for whom. When utilized in vague but authoritative ways, sustainability as a framework can end up redressing destructive sociopolitical systems in new "green" clothes (Checker 2011; Dooling 2009).

The above suggests that sustainability as a broadly articulated framework can miss particular nuances or context-specific issues in the places that it is applied (Friedman et al. 2014). For example, sustainability is unlikely to apply to different contexts (urban, suburban, exurban, rural) in the same way. This raises questions such as, if we know places are different, then why would we use the same language and tools to define sustainability? Despite these ongoing challenges, sustainability as a mission or goal has been taken up everywhere from corporations to cities to rural cooperatives.

In this chapter, we contribute to the ongoing effort to consider, evaluate, redefine, and apply sustainability. We contend that more common definitions of sustainability still maintain rather broad generalizations and, like any good framework, should be routinely reviewed and critiqued (Gibson 2006), particularly when applying it to specific contexts. Similarly, we believe a focus on the normative values that are interwoven with these definitions can help to better focus on equity and justice.

We assert that sustainability's flexible and broad conceptualization provides guidance and inspiration, but it must be developed through place-based experiences and nuance. In that spirit, our goal in this chapter is to draw on some of sustainability's competencies (Wiek, Withycombe, and Redman 2011) and lower order concepts as an example of how to take a broad, flexible definition and apply it to specific goals. To do so we consider the development and implementation of a graduate/undergraduate level course titled "Natural Resource Leadership" that focused on water resources, leadership, engagement, experience, equity, and service as a way to develop innovative conceptualizations of sustainability. In other words, we demonstrate how we used the experiential classroom to define sustainability.

Natural Resource Leadership (NRL) was offered in May 2015 at Chatham University in Pittsburgh, PA. A part of Chatham's nascent sustainability program, the NRL course explored the social, ecological, and economic aspects of water, and then applied sustainability to resource challenges. Field experiences were critical to the course goals, and included visits across rural, suburban, and urban sites, service projects, and diverse forms of engagement with both water and the communities that manage it. A key aspect of the course was whitewater kayaking, whereby students learned recreational skills as an additional way to engage with communities and resources in diverse ways (it is important to point out that one of the professors was a certified kayak instructor; any experiential component of a course should include the appropriate expertise). The course spanned three intensive weeks, meeting four days weekly, plus an overnight field trip during the final week. Course objectives or outcomes—the things students will be able to do or have learned once they have successfully taken the course—included an emphasis on conceptualizing sustainability, equity, and systems, and using leadership and engagement to solve human/nonhuman challenges.

Our approach to teaching includes the principle that universities should not educate in a single place and in silos (Vincent, Bunn, and Sloane 2013), but instead should imagine the university across a vast and diverse geography (Phillips et al. 2011). Additionally, we focus on critical practice and pedagogy in an attempt to query the inequities that emerge from environmental change (Jarosz 2004; Henderson and Zarger 2017).

We focused on water resource governance and sustainability in the Pittsburgh Metropolitan Region as well as the surrounding suburban and rural communities in part because of longstanding water challenges related to mining. Pittsburgh's location at the northern end of the Appalachian coal seam has meant the metropolitan area has historically served as a hub for coal and steel production. Coal and steel serve not only a historic economic role in the region, but also a cultural role (see, e.g., the National Football League team the Pittsburgh Steelers). While the majority of the manufacturing sector left the region decades ago, and coal production continues to decline, the legacy on the social and environmental landscape is enduring. Posited as part of the expansive "rust belt" region, much of that landscape bears the legacy of industrial manufacturing, particularly in the form of steel and coal, along with efforts to green these postindustrial landscapes (Finewood 2016; Schilling and Logan 2008).

Specifically, we attend to the issue of acid mine drainage (AMD) as a particular resource challenge. The region is home to a vast range of mine cavities—some known and some unknown—that, over the years, have stored water like aquifers. These mines slowly leak or leach water that has picked up contaminants left exposed from mining activities. The familiar image of bright orange oxidation marks rural, suburban, and urban landscapes. This presented an opportunistic way to engage with local communities and water resource challenges that also overlapped with other course objectives. Finally, like many other regions in the U.S., there is a renewed focus on sustainability in the form of greening, green spaces, renewable energy, and so forth. Thus the region presents an excellent opportunity to think about how different forms of sustainability are mobilized.

In the remainder of this chapter, we consider the many aspects of planning and implementing the NRL course. In the following section we discuss the different components of, and our strategy for defining, sustainability. In the penultimate section we briefly review the schedule in hopes that others can adopt strategies appropriate to their needs. We conclude with a discussion of why this matters.

Regional Sustainability Defined by Local Contexts and Learning Outcomes

In this chapter, we draw on the NRL course as a case study to expand the definition of sustainability. In this section we define the learning outcomes and experiences that link together diverse aspects of sustainability as applied in the course. As mentioned above, sustainability is probably most commonly defined as meeting the needs of the present without compromising the needs of future generations. This conceptualization makes sense as a general edict. In some form or another, societies generally want to utilize resources but also protect the needs of future generations. This acknowledges the human well-being/environmental health link the Brundtland architects were after. However, again, if sustainability conceptualizations are useful it is because they provide broadly positive advice. Action happens in the tasks.

When creating the course, our approach to applying and defining sustainability was through local contexts and student experiences. At the risk of overusing metaphors, we began by thinking about sustainability through a pedagogical conceptualization used to describe private property rights as a "bundle of sticks" (Freyfogle 2003). From this perspective, although the government generally protects private property rights, "ownership is not one aggregate right; it is many distinct rights, and a landowner can possess few or many of them" (Freyfogle 2003, 19). In other words, private property exists everywhere, but every place may contain a bundle of property rights that are different from other places in both kind and amount.

Drawing on the above, sustainability can be conceived of—and is often defined as—a broad goal. But, depending on the context, different objectives might apply to meet that goal. So from our perspective sustainability is both a broad, guiding framework but also a localized bundle of ideas, from which communities utilize those that appropriately apply to their contexts. For example, coastal property regulations are particularly concerned with flood zones or high-tide lines. These may be regulated heavily by state or federal agencies (and inform different approaches to sustainability). Sustainability would not be thought of the same way in suburban communities. Property law in suburban areas may pay particular attention to roads or lawn aesthetics, the latter of which could be regulated by homeowners' associations. Despite being contextually diverse, these can both be characterized as bundles of property rights. If we think of sustainability in this way, then it emphasizes the importance of utilizing the right tool for the specific community goals or needs. This can help prioritize budgets, desired outcomes, or specific skillsets relevant to the context at hand. This also points to the need for an expanded, flexible definition (Friedman et al. 2014). For us this meant, rather than thinking of sustainability as a three-legged stool (economic, social, environmental), we instead focused on ideas that we felt important and appropriate for the time and context.

With that in mind, we next turned to Wiek, Withycombe, and Redman's "competencies" (2011), or the skills that employers look for when hiring into the professional sustainability field. The authors present five key competencies that students should have once they graduate: systems-thinking, anticipatory, normative, strategic, and interpersonal. In our view, these conceptualizations begin to reflect an adequate definition of sustainability in action. We drew specifically on normative ("concepts of justice, fairness, responsibility, safety, happiness, etc."), interpersonal ("concepts of leadership"), and strategic ("adaptation and mitigation") competencies as high order framing (Wiek, Withycombe, and Redman 2011). We then shifted to thinking about this in the context of regional water resource management challenges.

We quickly felt that these high level competencies provide good guidance and help frame broad ideas, but need to be further defined for local action. Specifically, we note that ideas about experience and engagement were largely implicit aspects of broader competencies. Such implied concepts were much more vague and somewhat open for interpretation, despite being important for our ideas of sustainability. For example, a high order focus on "social systems" as a coequal part of sustainability is important, but can be irrelevant if it does not explicitly encompass the normative values of community members or democratic engagement (Chini et al. 2016). These latter concepts are how sustainability *actually gets done*. Thus we began with higher-order competencies but combined them with lower-order concepts that helped shape our unique approach to sustainability within the course: leadership, experience, and engagement with regional water resource challenges.

Next, we thought again about the regional challenges of managing the ongoing waste stream generated from historic extraction economies. AMD typically results from disturbances due to mining or other, similar activities. When these sites contain an abundance of sulfide minerals (as is the case for coal extraction and handling sites) water outflows often contain toxic metals and are highly acidic. This liquid, or AMD, can have detrimental impacts on local rivers and streams. While much work has been done to mitigate and adapt to the AMD problem, it is persistent and in need of ongoing innovation for solutions. Sustainability provides a key framework.

Finally, the learning outcomes (Table 13.1), or what students should be able to do after they successfully take the class, served as targets toward which we mobilized the sustainability competencies and topics. In other words, competencies, topics, and learning outcomes were our unique bundle of sticks. Learning outcomes help to determine ways to instrumentalize the ideas (e.g., kayaking as a way to actually get students on the water). These outcomes also connect with the mission of the sustainability program at Chatham University.

Table 13.1. Sustainability program objectives in the Natural Resource Leadership (NRL) course

---

Conceptualizing Sustainability—Students will be able to explain the origins, meanings, and applications of sustainability, and by extension, explain the interrelationships among environmental, societal, and economic wellbeing. They will do this in a framework that recognizes the cultural dimensions of sustainability.

Systems Thinking—Students will develop tools to model complex systems, describe the impact of changes within systems, consider the impacts of decision-making on systems, and analyze a system's strengths and weaknesses.

Transformative Leadership—Students will be prepared to take an active role in advancing sustainability, with the understanding that to do so will require behavioral, cultural, institutional, and other changes at multiple spatial and temporal scales.

Creativity—Students will understand that facilitating sustainable attitudes and practices requires creativity in conceptualizing existing conditions and generating and implementing sustainable solutions to complex problems.

In sum, our course goal was to define and apply sustainability through the NRL course. We considered the classroom as a way to get out in the field and engage. Following a discussion of specific learning outcomes (immediately below), in the next section we describe our course and the ways we applied our conceptualization of sustainability to pedagogical practice.

Learning Outcome: Conceptualizing Sustainability

The first learning outcome for the course was *Conceptualizing Sustainability*, where students will be able to define, explain, and apply sustainability. Here students who participate, engage in, and do the work for the course should be able to think through ideas about sustainability in different ways and as they might apply in diverse local contexts. As noted earlier, although water issues are certainly diverse in the region, we principally connected to the history and impacts of Appalachian coal mining. Broadly speaking we compelled students to consider what sustainability actually means when applied to human/nonhuman relationships such as these.

The region's historic network of mines, hydrology, land use, and culture represent a socioecological system that results in a specific water pollution issue: AMD. An important question we posed to class participants was, *how and why was coal and steel production sustained despite some of its negative social and environmental costs?* Here we pushed students to explore the role of culture and economics in support of these industries. We also compelled students to consider different forms of adaptation and mitigation strategies for solving these problems, and how sustainable they would be over time. In this sense we asked, *what, how, and why is something sustained? What local contexts shape these strategies and outcomes?*

Learning Outcome: Systems Thinking

The second learning outcome was *Systems Thinking*, where students would develop tools to understand both systems and the interactions of humans and nonhuman nature within them. Here we primarily attend to ideas from both ecology (Walker and Salt 2006) and political ecology (Ranganathan 2015; Angelo and Wachsmuth 2014). Ecology, on the one hand, helps us to think about systems and how the parts interact. Political ecology, on the other hand, considers the struggles, power dynamics, and inequalities that happen within those systems.

Specifically, we integrated concerns about equity and justice, or fairness in term of participation and outcome, which links to normative and strategic competencies. Watersheds and systems provide a clear setting to think about systems but then explore struggles for equity and justice as a result of environmental change and decision-making (Walker and Bulkeley 2006). Importantly, we considered ideas about spatial equity, not as a coequal "leg on the stool," but as a concern that is weaved throughout the process (Holifield, Chakraborty, and Walker 2018).

Further linking to normative and strategic competencies, we feel that student experiences should be in partnership with the community, asking how we can collaborate and solve particular challenges, as opposed to bringing preconceived ideas to the community. To design this aspect of the course content and field experiences we responded to the challenges of student learning set forth by *A Crucible Moment* (National Task Force on Civic Learning and Democratic Engagement 2012):

- demands for institutional local and global relevance and a focus on life's big questions,
- highly engaged in community-based partnerships and social change,
- represent the wide diversity of demographic and geographical communities, and
- community service as alternative form of political engagement.

Our partnership with Friends of the Cheat (cheat.org), for example, was central to the design. They were engaged from the outset of course planning in the spirit of true community engagement: planning with our partner, not for the partner. As a result, service components of the course had a greater concern for equity as expressed by community members, rather than coming directly from professors. Our engagement with engineers, park managers, and watershed groups led to an understanding of the issues, challenges, and strategies that communities must work with. We were able to explore what equity issues actually exist in the region, such as the difference between urban and rural poverty. Thinking about the political ecology of systems obliges students to ask, *how are the cost and benefits of environmental change unevenly distributed?*

Learning Outcome: Transformative Leadership

The third learning outcome was *Transformative Leadership*, where students take an active role in participating in and advancing sustainability. Here we return to higher-order interpersonal and strategic competencies. We combined these with lower-order concepts such as service, engagement, experience, and leadership. In particular, we see pedagogical value for including ideas about leadership and engagement in a sustainability course. As educators we look for unique pedagogies that not only focus on sustainability, but also promote a new lens on future education strategies. The National Task Force on Civic Learning and Democratic Engagement (2012), for example, finds that civic learning in traditional academic disciplines is complementary rather than competitive. The report further points to research findings that suggest that students who participate in civic learning opportunities are more likely to persist in college and complete their degrees, obtain skills prized by employers, and develop habits of social responsibility and civic participation.

> The beneficiaries of investing in CLDE are not just students or higher education itself; the more civic-oriented that college and universities become, the greater their overall capacity to spur local and global economic vitality, social and political well-being, and collective action to address public problems. (National Task Force on Civic Learning and Democratic Engagement 2012, 2)

A challenge for higher education institutions is to "delineate multiple educational pathways in the curriculum and cocurriculum—appropriate to institutional mission and fields of study—that incorporate civic questions, pedagogies, and practices for all students" (National Task Force on Civic Learning and Democratic Engagement 2012, 32). We paid attention to this by asking students, *how do we engage, learn from, and (when appropriate) lead on sustainability challenges?*

Learning Outcome: Creativity

The final learning outcome was *Creativity*, where students will recognize that engaging with these topics and contexts requires a wide range of strategies, including those that may be considered outside common practice. In this case, we considered the importance of engagement with both human caused problems and the nonhuman nature through which we work. In other words, we got on the water to get students' boots muddy.

We put this to work in three unique ways. First, we developed three field trips that would have students interacting with AMD in diverse land use contexts: urban, suburban, and rural (more details on this below). At these sites we engaged with the engineers, managers, and advocacy organizations that manage these sites and are responsible for both their function and their justification (cost, etc.) in the public eye. Second, we built a specific service aspect into one of our site visits. Working with Friends of the Cheat we undertook a restoration site clean up as well as other river sweeps. These were coupled with several visits to different problem spots.

Lastly, we incorporated whitewater kayak training into the course as a way to physically engage with water. We wanted students to experience and engage in multiple, unique ways. Students were given basic kayak training in a safe environment (Chatham's indoor pool) and then an on-the-water paddle across the urban Pine Creek, which flows from the northern suburbs into the Alleghany River. Finally, students kayaked class II-III whitewater at the Cheat River. This was undoubtedly the centerpiece of the course that connected all of the disparate parts. While on both the urban and rural rivers, for example, students were able to actually see and touch the issues we had been discussing. It also presented a new way to ask the question, *how do we find multiple, creative ways to engage with our topic and context?*

NRL: Sustainability Defined as Leadership, Experience, and Engagement

In this section we outline our schedule for implementing the NRL course. As stated above, our primary goal for the course was to explore leadership, experience, and engagement with water resource management through a sustainability lens. However, broad, conventional definitions felt too vague or nebulous to apply to an experiential class. We began instead by drawing on specific sustainability competencies, topics, and outcomes that we felt valuable to what we wanted to do in the class.

As noted previously, our approach to teaching meant it was particularly important to engage in different suburban, urban, and rural contexts, and to view that engagement as more than just transactional. This is, in our view, the difference between showing up and truly engaging. Both classroom and experiential learning provided these opportunities, and creative engagement made them unique and compelling. Finally, we wanted students to think through what leadership actually means in the context of regional sustainability challenges and how it can create sustainable action.

Below we discuss the class structure and outcomes through summaries of a three-week, intensive schedule. Weekly agendas were meant to be cumulative, whereby students develop skills that strengthen learning and engagement in each successive meeting. In the final section of this chapter we conclude with a brief discussion of the values we believe were created through the class and how others might take on such a course.

Week One: In the Classroom, in the Pool

In week one, we focused on kayak training and classroom experience. We spent several contact hours in the pool teaching students basic whitewater skills. In addition, we merged these skills with readings and lectures on water quality, water policy, pollution, and leadership (Kouzes and Posner 2007; Hoornbeek 2011). Classroom contact hours were more traditional with faculty lecturing on key issues, prompting students with questions, and creating discussion around kayaking skills development, leadership, and watershed issues. In this time period students also learned about AMD and its regional impacts.

Week Two: Getting Out into the Field

In week two, we started getting out into the field. We spent day one reviewing concepts—including sustainability, water quality issues, leadership, and regional challenges, coupled with at least one hour in the pool. On the second, third, and fourth days we made several field visits. We visited two restoration sites. The first was Emerald View Park (mwcdc.org/park), a newly created urban green space with AMD issues. We spoke with the site manager about how they were incorporating research and monitoring in their management plan. On the next day we visited Wingfield Pines (alleghenylandtrust.org/green-space/wingfield-pines), a suburban restoration site created to manage AMD from draining into Chartiers Creek. There we spoke with one of the design engineers about their challenges and strategies for managing the AMD source. On the final day of week three we kayaked Pine Creek, a class I urban stream. This provided an opportunity for students to hone their kayaking skills and see the city from the water's point of view.

Week Three: Engagement

In the final week, we traveled to the Cheat River in West Virginia to camp, kayak, and work with the Friends of the Cheat. Over three days the class focused on AMD in a rural area. During the evenings we proctored "campsite" modules about leadership as well as comparisons of AMD across rural-suburban-urban contexts. We also kayaked the Cheat River Narrows, a class II-III section with several opportunities for students to apply their newly developed skills as well as see from the river's point of view. Faculty and students also participated in three service projects, cleaning up restoration sites and campsites owned by Friends of the Cheat. During each service project, the participants learned from the executive director about the particular challenges of AMD adjacent to popular rural recreation sites.

Conclusion

The Natural Resource Leadership course was developed as a way to merge competencies, concepts, and experiences into a definition of sustainability that can be applied to specific water resource challenges. We considered ideas about leadership, experience, and engagement, and how these concepts could help students think about ways to work toward sustainable solutions for vexing problems. In participating in the class, students developed competencies, received kayak training, and actualized sustainability strategies for AMD.

Since the course only ran once we do not have significant outcome assessments. However, we think there are opportunities to think about our strategy as a way to continue defining sustainability across diverse contexts. Student responses were, to put it colloquially, off the charts. They relished in the opportunity to learn concepts and engage with issues in multiple, creative ways. Students often reflected on the challenges of defining sustainability, in finding purchase as a leader, and the importance of engagement to their course experience. As expected, students particularly enjoyed whitewater kayaking as a way to see from the resources' point of view.

If possible, we highly encourage other instructors to find creative ways to define and apply sustainability in their classrooms. Moving beyond traditional approaches took some convincing for both students and administration, but the outcome was worth it. Further, as threats to U.S. water systems continue to expand (Gleick 2016, 555), it becomes incumbent on educators to find new ways to implement sustainability strategies in and out of the classroom.

References Cited

Allouche, Jeremy. 2011. "The Sustainability and Resilience of Global Water and Food Systems: Political Analysis of the Interplay Between Security, Resource Scarcity, Political Systems and Global Trade." *Food Policy* 36: 3–8.

Angelo, Hillary, and David Wachsmuth. 2014. "Urbanizing Urban Political Ecology: A

Critique of Methodological Cityism." *International Journal of Urban and Regional Research* 39(1): 16–27.

Carr, Edward R., Michael H. Finewood. 2015. "Sustainable Development." In *Global Environmental Problems: Causes, Consequences, and Potential Solutions*, edited by Tobias J. Lanz, 215–225. California: Cognella Academic Press.

Checker, Michelle. 2011. "Wiped Out by the 'Greenwave': Environmental Gentrification and the Paradoxical Politics of Urban Sustainability." *City & Society* 23(2): 210–229.

Chini, Christopher M., James F. Canning, Kelsey L. Schreiber, Joshua M. Peschel, and Ashlynn S. Stillwell. 2017. "The Green Experiment: Cities, Green Stormwater Infrastructure, and Sustainability." *Sustainability* 9(1): 105.

Dooling, Sarah. 2009. "Ecological Gentrification: A Research Agenda Exploring Justice in the City." *International Journal of Urban and Regional Research* 33(3): 621–639.

Finewood, Michael H. 2016. "Green Infrastructure, Grey Epistemologies, and the Urban Political Ecology of Pittsburgh's Water Governance." *Antipode* 48(4): 1000–1021.

Freyfogle, Eric T. 2003. *The Land We Share: Private Property and the Common Good*. Washington: Island Press.

Friedman, David S., Yehuda Klein, Jose Pillich, and Michael T. Sullivan. 2014. "Sustainability Planning, Environmental Justice and Climate Change: Applications of the Long Island MARKAL Model." *Suburban Sustainability* 2(2): 1–23.

Gibson, Robert B. 2006. "Sustainability Assessment: Basic Components of a Practical Approach." *Impact Assessment and Project Appraisal* 24(3): 170–182.

Gleick, Peter H. 2016. "Water Strategies for the Next Administration." *Science* 354(6312): 555–556.

Henderson, Joseph A., and Rebecca K. Zarger. 2017. "Toward Political Ecologies of Environmental Education." *Journal of Environmental Education* 48(4): 285–289.

Heynen, Nik, Harold A. Perkins, and Parama Roy. 2006. "The Political Ecology of Uneven Green Space: The impact of Political Economy on Race and Ethnicity in Producing Environmental Inequality in Milwaukee." *Urban Affairs Review* 42(1): 3–25.

Holifield, Ryan, Jayajit Chakraborty, and Gordon Walker, eds. 2018. *The Routledge Handbook of Environmental Justice*. New York: Routledge.

Hoornbeek, John A. 2011. *Water Pollution Policies and the American States: Runaway Bureaucracies or Congressional Control?* New York: SUNY Press.

Jarosz, Lucy. 2004. "Political Ecology as Ethical Practice." *Political Geography* 23: 917–927.

Kouzes, James M., Barry Z. Posner. 2007. *The Leadership Challenge*, 4th ed. San Francisco: Wiley and Sons. Meadows, Donella H. 2008. *Thinking in Systems*. White River, VT: Chelsea Green.

National Task Force on Civic Learning and Democratic Engagement. 2012. *A Crucible Moment: College Learning and Democracy's Future*. Washington, D.C.: Association of American Colleges and Universities.

Phillips, J., A. Benjamin, R. Bishop, L. Shiqiao, E. Lorenz, L. Xiaodu, M. Yan. 2011. "The 20-Kilometer University: Knowledge as Infrastructure." *Theory, Culture & Society* 7–8: 287–320.

Ranganathan, Malini. 2015. "Storm Drains as Assemblages: The Political Ecology of Flood Risk in Post-Colonial Bangalore." *Antipode* 47(5): 1300–1320.

Redman, Charles. 2014. "Should Sustainability and Resilience Be Combined or Remain Distinct Pursuits?" *Ecology and Society* 19(2): 37.

Schilling, Joseph, and Jonathan Logan. 2008. "Greening the Rust Belt: A Green Infrastructure Model for Right Sizing America's Shrinking Cities." *Journal of the American Planning Association* 74(4): 451–466.

Vercoe, Richard, and Robert Brinkmann. 2010. "A Tale of Two Sustainabilities: Comparing Sustainability in the Global North and South to Uncover Meaning for Educators." *Education.*

Vincent, Shirley, Stevenson Bunn, and Lilah Sloane. 2013. *Interdisciplinary Environmental and Sustainability Education on the Nation's Campuses 2012: Curriculum Design.* National Council for Science and the Environment, Washington, D.C.

Walker, Brian, and David Salt. 2006. *Resilience Thinking: Sustaining Ecosystems and People in a Changing World.* Washington, D.C.: Island Press.

Walker, Gordon P., and Harriet Bulkeley. 2006. "Geographies of Environmental Justice." *Geoforum* 37(5): 655–659.

Wiek, Arnim, Lauren Withycombe, and Charles L. Redman. 2011. "Key Competencies in Sustainability: A Reference Framework for Academic Program Development." *Sustainability Science* 6(2): 203–218.

World Commission on Environment and Development. 1987. *Our Common Future.* Oxford University Press: New York.

14

Managing Wildlife amid Development

A Case Study of Sustaining
Mountain Lion Populations in California

MELISSA M. GRIGIONE, MICHAELA C. PETERSON,
RONALD SARNO, AND MIKE JOHNSON

Habitat loss and degradation are currently the greatest threats to global biodiversity—especially that of large, wide-ranging carnivores (Purvis et al. 2000; Ripple et al. 2014; Schipper et al. 2008). For the increasing number of wildlife populations living in semideveloped landscapes, action is required to assess whether and how sufficient habitat can be preserved within a matrix of suburban and exurban development (Burdett et al. 2010; Fernandez-Juricic and Jokimaki 2001). Conserving apex predators in suburban systems is particularly desirable, as they are vital to normal ecosystem functioning (Beschta and Ripple 2009; Estes et al. 2011; Terborgh et al. 2001). Not only is ecosystem function important, but the ramifications of absent predators and fluctuating predator densities can impact human health. For example, increases in Lyme disease in the northeast and midwestern U.S. over the last 30 years are not correlated with deer abundance, but with the range-wide decline of the red fox (*Vulpes vulpes*) an important small-mammal predator, due to the range expansion of coyotes (*Canis latrans*); and coyotes have expanded their range as a result of the large-scale decline of wolves (*Canis lupus*) by humans. Furthermore, across Wisconsin, Pennsylvania, New York, and Virginia there is low spatial correlation between deer abundance and the incidence of Lyme disease. However, it appears that coyote and red fox abundance accurately predict the spatial distribution of Lyme disease in New York (Levi et al. 2012). These results highlight the importance of apex predators in urban and suburban ecosystems. Changes in predator abundance and/or community structure can initiate trophic cascades that ultimately influence human health.

A major concern for wildlife occupying suburban habitats is whether such habitats act as sinks—where mortality is higher than survival and populations can only be sustained through immigration from more optimal habitats (Riley, Gehrt, and Cypher 2010; Vierling 2000). Urban regions in the northern Sierra Nevada Mountains in California, for instance, act as sinks for black bears (*Ursus americanus*) because of high mortality, which is not offset by age-specific fecundity (Beckmann and Lackey 2008). However, not all urban habitats act as sinks for carnivores. For example, a thriving population of the endangered San Joaquin kit fox (*Vulpes macrotis mutica*) exists in highly urbanized Bakersfield, California (Cypher 2010). Many species, such as spotted hyenas (*Crocuta crocuta*) and golden jackals (*Canis aureus*), persist in developed areas through scavenging and occasional predation on livestock (Abay et al. 2011; Cirovic, Penezic, and Krofel 2016). Even large carnivores such as tigers (*Panthera tigris*) can persist in areas of high human density (Carter et al. 2012). In general, the viability of (sub)urban carnivores depends on local prey abundance (Abay et al. 2011; Kertson et al. 2011), connectivity between fragments of natural habitat (Fernandez-Juricic and Jokimaki 2001; Riley et al. 2014), amount of mortality associated with vehicle collisions (Andrews, Gibbons, and Jochimsen 2008; Beckmann and Lackey 2008), and human tolerance (Abay et al. 2011).

Case Study

The challenges of wildlife management in developed landscapes can be highlighted by examining mountain lions (*Puma concolor*) in California. Over 50 percent of land in California is privately owned, and undeveloped habitat is highly fragmented and bisected by numerous roads and interstates (Thorne, Cameron, and Quinn 2006). In order for mountain lions to persist across the state, regional-level conservation plans (Beier 1993) that incorporate developed lands, while integrating projected growth estimates for human populations (Burdett et al. 2010), are required.

Mountain lions exhibit considerable variation in movement; some individuals use highly developed areas extensively (Beier et al. 2010; Burdett et al. 2010), while most generally avoid anthropogenic habitat features within their home ranges (Belden and McCown 1996; Burdett et al. 2010; Smallwood 1994; Spreadbury et al. 1996; Van Dyke et al. 1986), or shift peak activity between sunset and sunrise when close to humans (Van Dyke 1984). Individual mountain lions that avoid even moderate development have lower mortality than those who use developed areas in proportion to availability (Burdett et al. 2010). Habitat models should therefore include developed areas as potential conduits between fragments of natural habitat for mountain lions, which may be utilized by mountain lions for dispersal or migration, while recognizing that such areas are generally not sufficient to support resident mountain lions.

By comparing home range size and overlap across three ecologically distinct regions of California, we are able to develop a deeper understanding of variation in mountain lion habitat utilization amidst development. Illuminating the extent of variation in ranging behavior will enable managers to make more enlightened decisions regarding designation of priority areas for conservation as well as the placement and design of corridors that will facilitate movements between important habitats.

Data for these analyses—obtained from 56 mountain lions—were collected over a period of 10 years (from 1983 to 1993). While dated, these data provide a unique opportunity to explore the influences of natural and anthropomorphic features on mountain lions throughout diverse regions of California. While the data provide us with only a snapshot of time (subsequent data have not been collected on these populations), we believe that the results transcend time because they underscore crucial aspects of mountain lion spatial ecology that are, indeed, relevant today.

STUDY SITES

Data for this study were collected from three study sites in California (Figure 14.1). The study areas include the Diablo Range (parts of Santa Clara, Merced, Stanislaus, San Joaquin, and Alameda Counties), the Santa Ana Mountains (parts of Los Angeles, Orange, San Bernardino, Riverside, and San Diego Counties), and the central Sierra Nevada Mountains (eastern Fresno County).

The study sites listed above comprise two distinct ecosystems: a low-elevation, chaparral-oak woodland Coastal Range ecosystem (Diablo Range and Santa Ana Mountains), and a high-elevation mixed-conifer ecosystem (Sierra Nevada). In contrast to the Sierra Nevada (SN), seasonality is not pronounced in the Coastal Ranges (CR) and snowfall is uncommon. Black-tailed deer (*Odocoileus hemionus*), the primary prey species of mountain lions in all three study areas, are not migratory in the Coastal Range, as they are in the Sierra Nevada.

Diablo Range

The Diablo Range is roughly 20 km east of San Jose, California (Figure 14.1). The Mount Hamilton area of the Diablo Range, consisting of 550 km² of public and private lands, was the primary study area. However, 1,300 km² of aerial monitoring included areas adjacent to this core area. The habitat is within the inner coast of California and consists of oak (*Quercus* spp.) woodlands and chaparral. Elevation is from 300 to 1,100 m, mostly above 600 m. Major ridges lie in a NW-SE direction. Water is available year-round from springs and cattle ponds. Winters are wet and cool with temperatures ranging from 12°C to 20°C. Summers are dry and warm with temperatures from 18°C to 31°C. Snowfall is uncommon and rainfall is highly variable with an accumulation of 30–87 cm. Nonconsumptive recreation is the primary activity on these public lands, while sport hunting and cattle grazing are the primary activities on private lands.

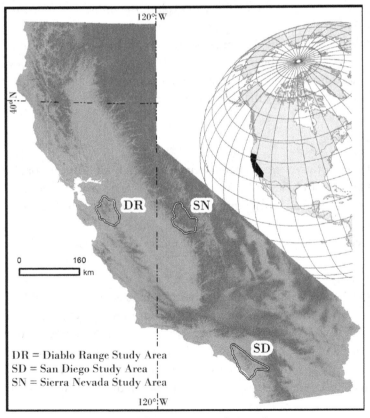

Figure 14.1. California mountain lion study areas.

Santa Ana Mountains

From January 1987 to August 1989 data were collected in the Camp Pendleton area and surrounding lands, including the southern half of the Santa Ana Mountains and adjacent foothills (Figure 14.1; Padley 1990). The study area was 1,350 km² and bounded by three highways and the Pacific Ocean. Activities on the study site include military training, farming, and cattle ranching. The climate is cool with wet winters (October–April) and hot, dry summers (May–September). Mean winter and summer temperatures are 24°C–26°C and 26°C–28°C, respectively. Mean annual precipitation is 34.5 cm. Snow is rare, but fog is common during the wet season. Elevations range from sea level to 998 m with major drainages having locally intermittent flows running from east to west. Oak (*Quercus* spp.) woodlands dominate canyon bottoms and on rolling topography in areas away from the coast. Riparian areas contain willow (*Salix* spp.), alder (*Alnus rhombifolia*), and sycamore (*Platanus racemose*). Coastal slopes are characterized by sage and scrub communities. Grassland communities are found throughout the study area, and orchards are located east of Camp Pendleton, a military base.

An additional study was conducted in the Santa Ana Mountains from April 1988 to February 1993 (Beier and Barrett 1993). The study site encompassed 2,070 km² in the Santa Ana Mountains and, during the last 2.5 years of the investigation, included the Camp Pendleton area (Figure 14.1). Sixty-one percent of the study area is in public ownership (Forest Service, Bureau of Land Management, Camp Pendleton, Naval Weapons Station, state parks, county parks) and private reserves. Study area elevations range from sea level to 1,690 m. The higher elevation conifer forests include bigcone Douglas fir (*Pseudotsuga macrocarpa*), Coulter pine (*Pinus coulteri*), and cypress (*Cupressus forbesii*).

Sierra Nevada

This study area encompassed 2,072 km² in eastern Fresno County, mostly within the Sierra National Forest (Figure 14.1), ranging from 244 to 4,000 m in elevation. Activities on this land include recreation, logging, and livestock grazing. Black-tailed deer density was 0.868 deer/km². Oak (*Quercus* spp.) woodland and yellow pine (*Pinus jeffreyi*) forests are located in lower elevation areas (below 1,200 m) and yellow pine, white fir (*Abies concolor*), mixed conifer, and red fir (*Abies magnifica*) forests are located above 1,700

m. These forest types correspond to winter and summer ranges, respectively, for migrating black-tailed deer and mountain lions.

METHODS

Mountain lion radiotelemetry data were obtained from four populations. Telemetry data were collected from 1983 to 1993. Two separate studies (1986–1989 and 1988–1993) were conducted within the same general area in the Santa Ana Mountains, and monitoring of many individuals extended from the first study into the second.

Diablo Range

In the Diablo Range, radiotelemetry data were collected from March 1984 to June 1989 on 24 mountain lions (Hopkins 1989). Aerial radiotelemetry readings occurred weekly, with no more than one location/day/lion used for HR analyses. Black-tailed deer density was 10.5 deer/km². The ratio of black-tailed deer to mountain lions was 553:1 (proportional method) and 210–350:1 (aerial method; Hopkins, Kutilek, and Shreve 1986; Hopkins 1989).

Santa Ana Mountains

Radiotelemetry data were collected from January 1987 to August 1989 in the Camp Pendleton area on eight mountain lions (Padley 1990). Radiotelemetry readings were obtained daily from the ground and two to four times per month from an airplane. Mean length of monitoring was 20 months/animal. Beier and Barret (1993) collected radiotelemetry data on 32 mountain lions from April 1988 to February 1993. Radiotelemetry readings were obtained every 1–4 days from the ground and every 10 days from an airplane. Helicopter surveys indicated 5 black-tailed deer/km² in good habitat and 1.9–2.3 black-tailed deer/km² in poorer habitats.

Sierra Nevada Mountains

Radiotelemetry data in the Sierra Nevada were collected from August 1983 to December 1986 on 16 mountain lions (Neal 1990). Mountain lions were monitored daily from the ground using radiotelemetry for 0.1–4.5 years. Black-tailed deer density in the study areas was estimated at 0.868 deer/km².

We used Universal Transverse Mercator Grid coordinates (UTMs) to create home ranges (HRs). The CALHOME HR analysis program, MS-DOS version 1.0, was used to calculate 95 percent minimum convex polygons for each mountain lion, during each season (winter and summer) and year

(CALHOME 1994). Home ranges were exported to ArcInfo 7.1.2 for Windows NT where ecological coverages were intersected. GIS coverages for river (River Reach File 3alpha, 1: 100,000 scale, units=meters), road (USGS DLG-3, 1:100,000 scale, units=meters), and vegetation data (California Gap Analysis Vegetation: Wildlife Habitat Relations, 100 ha minimum mapping unit for upland vegetation and 40 ha minimum mapping unit for wetland vegetation) were intersected with HRs to determine the percentage and actual amount of each landscape feature in HRs (i.e., exclusive and overlapped portions). In each study area trapping efforts were intensive enough to ensure that almost all mountain lions were collared, so that calculations of overlapped vs. exclusive HR would be biologically meaningful. A total of 60 vegetation classes were used for this analysis. River data represent all blue lines on USGS 1:100,000 quadrangle maps (including channels, shorelines, and ephemeral streams). Road data do not include private roads, U.S. Forest Service roads, logging roads, or trails on public land.

For each season/year, we determined the size of each HR (km^2) and the amount of overlap between HRs (km^2). The amount of overlap was measured in two ways: (1) as area of overlap and (2) as percentage overlap of the entire HR. In addition, we broke overlap down by sex to explore how males and females differed with respect to HR overlap.

Deer densities had been calculated previously for each study area (Beier and Barrett 1993; Hopkins 1989; Neal 1990; Padley 1990). Since different methods were used to determine densities, these estimates were not incorporated into any of our statistical analyses. However, the known differences in deer density and distribution were used to inform data interpretation. All statistical analyses were performed using the Number Cruncher Statistical System (NCSS 1996).

HOME RANGE SIZE AND OVERLAP

Home range size and overlap were transformed using natural logarithms and square roots, respectively. One-Way ANOVAs (alpha=0.05) were used for each study to determine if significant differences existed (by gender) between seasons and years with respect to HR size and overlap. Repeated measures ANOVAs (general linear model; alpha=0.05) were used to simultaneously analyze the effects of both season and sex on HR size and overlap. To detect differences in HR size and overlap amongst studies, one-way ANOVAs (alpha=0.05) were conducted for each season and sex. For all tests, we used average HR size (calculated for the entire period an animal was monitored) for each mountain lion, in order to avoid lack of independence among years.

ECOLOGICAL AND ANTHROPOGENIC VARIABLES

To evaluate differences in ecological and anthropogenic variables between exclusive and shared parts of the HR, GIS data were collected for every mountain lion HR. GIS data that were not normally distributed were sqrt(arcsin) transformed. One-way ANOVAs were run for each ecological and anthropogenic variable for each study site, season, year, and sex in order to test for differences between exclusive and nonexclusive portions of HRs. Alpha values were adjusted (Bonferroni's adjustment) for each study depending upon the number of years included in the analysis.

For vegetation data, we compared the number of habitat types in exclusive portions of the HR to nonexclusive portions. We did not use actual areas of each vegetation type because of large minimum mapping units (100 hectares=smallest polygon). Nonexclusive portions of HRs were divided further into areas shared by same sex individuals and opposite sex individuals in order to detect differences between intrasexual and intersexual interactions.

Rather than comparing actual units (m²) of river in exclusive and nonexclusive portions of the HR, we compared the density (m/m²) of river in exclusive and nonexclusive portions. We did this to standardize our riverine data and to obtain a more detailed understanding of the extent of river compared to other ecological variables in each HR. Using one-way ANOVAs, we compared the proportion of river in exclusive HR and nonexclusive portions for each study, year, season, and sex. Nonexclusive portions of HRs were subdivided into areas shared by same sex individuals and opposite sex individuals. Similar to river data, we used the density of roads (m/m²) in each HR to compare exclusive to nonexclusive portions. Nonexclusive portions were subdivided into areas shared by the same and opposite sex.

RESULTS

HR Size and Overlap

Home ranges were computed for 56 mountain lions. Average HR size and overlap for all studies are listed (Table 14.1). Male mountain lions in all studies have larger HRs than females for both seasons (Table 14.1). On average,

Table 14.1. California mountain lion home ranges by sex and season

Home Range Variables	Sierra Nevada Mountains	Diablo Range	Santa Ana Mountains
HOME RANGE SIZE (KM²)			
Male-winter	313 (+ 127)	105 (+ 16.1)	271 (+ 48.1)
Male-summer	258 (+ 39)	178 (+ 40)	287 (+ 56.3)
Female-winter	89 (+ 12)	53 (+ 7.7)	96 (+ 10.4)
Female-summer	167 (+ 26.2)	36 (+ 4)	101 (+ 18.9)
HOME RANGE OVERLAP (KM²)			
Male	31 (26% of total home range)	13 (26% of total home range)	39 (25% of total home range)
Female	32 (21% of total home range)	8 (12% of total home range)	31 (24% of total home range)

Notes: Average home range size (km²) and home range overlap (km²) for mountain lions in Sierra Nevada (1983–1987), Diablo (1984–1989), and Santa Ana (1987–1992) study sites, based on sex (male-female) and seasonal (winter-summer) differences.

male mountain lions across all studies share 25 percent of their HRs with other mountain lions. Females share 24 percent, 21 percent, and 12 percent of their HRs with other mountain lions in the Santa Ana, Sierra Nevada, and Diablo Ranges, respectively (Table 14.1).

ECOLOGICAL AND ANTHROPOGENIC FEATURES

One-way ANOVAs indicate that road density was significantly different between exclusive and shared portions of HRs at all three sites (Table 14.2). Lastly, in the Sierra Nevada during two seasons there was a greater density of roads in areas shared by males and females ($p=0.001$; $p=0.009$). In the Diablo Range road density was greater in exclusive portions of the HR during four seasons ($p=0.001$; $p=0.001$; $p=0.003$; $p=0.007$). In the Santa Ana, during two seasons there was a greater density of roads in exclusive areas ($p<0.004$), whereas during one season there was a greater proportion of roads in areas shared between males ($p=0.000$).

There were significant differences with respect to river densities in exclusive and shared portions of the HR in only the Diablo Range and the Santa Ana Mountains (Table 14.3). Specifically, in the Diablo Range, one-way ANOVAs show significant differences during seven seasons. Six of these seasons show a greater proportion of rivers in exclusive portions of the HR than in shared portions ($p=0.000$; $p=0.000$; $p=0.001$; $p=0.001$; $p=0.003$; $p=0.003$), while one season shows the opposite ($p=0.000$). In the Santa Ana,

Table 14.2. Significance of differences in road density between shared and exclusive home ranges

DIABLO RANGE	1984	1985	1986	1987	1988	1989
Male-winter	0.065	0.001*	0.003*	0.243	0.931	
Male-summer	0.261	0.443	0.089	0.188	0.056	
Female-winter	0.173	0.007*	0.929	0.394	0.086	
Female-summer	0.016	0.006*	0.025	0.164	0.474	0.001*
SANTA ANA	1987	1988	1989	1990	1991	1992
Male-winter				0.234	0.25	0.009
Male-summer				0.719	0.041	0.101
Female-winter	0.137	0.004*	0.431	0.248	< 0.001*	0.213
Female-summer	0.029	< 0.001*	0.02	0.828	0.115	0.074
SIERRA NEVADA	1983	1984	1985	1986	1987	
Male-winter			0.001*			
Male-summer			0.275			
Female-winter	0.193	0.884	0.42	0.738	0.298	
Female-summer		0.301	0.009*	0.439	0.183	

Notes: *P-value < 0.05, indicating a significant difference.
P-values associated with one-way ANOVAs comparing road density (m/m^2) between exclusive and shared portions of mountain lion home ranges for each study area (Diablo 1984–1989, Santa Ana 1987–1992, Sierra Nevada (1983–1987), based on sex (male-female) and seasonal (winter-summer) differences.

five seasons have significant differences in the proportion of rivers in exclusive and shared portions of female HRs. In three seasons there is a greater proportion of rivers in exclusive areas ($p=0.000$) whereas the other two seasons (both winter and summer 1991) have a greater proportion of rivers in areas shared with males ($p<0.0015$).

Similarly, there were significant differences in the number of habitat types in exclusive versus shared portions of HRs in only the Diablo Range and Santa Ana Mountains (Table 14.4). In the Diablo Range four seasons show a greater number of habitat types in exclusive portions of the HR ($p=0.000$; $p=0.003$; $p=0.003$; $p=0.007$), while one season contained a greater number of habitat types in shared portions of the HR ($p=0.001$). In the Santa Ana Mountains, six seasons show significant differences between habitat types in exclusive versus shared portions of HRs. In three of these seasons females have a greater number of habitat types in exclusive areas ($p=0.000$). In two seasons, females have a greater number of habitat types in areas shared with males than in exclusive areas or areas shared with other females ($p=0.000$). During the remaining season, exclusive portions of male HRs have more habitat types than areas shared with other males or females ($p=0.003$).

Table 14.3. Significance of differences in river density between shared and exclusive home ranges

DIABLO RANGE	1984	1985	1986	1987	1988	1989
Male-winter	0.589	0.001*	0.003*	0.231	0.186	
Male-summer	0.563	0.601	0.411	0.222	0.494	
Female-winter	0.216	0.003*	0.16	0.000*	0.001*	
Female-summer	< 0.001*	0.154	0.494	0.149	0.089	< 0.001*
SANTA ANA	1987	1988	1989	1990	1991	1992
Male-winter				0.578	0.723	0.284
Male-summer				0.085	0.362	0.216
Female-winter	0.071	< 0.001*	0.921	0.649	< 0.001*	0.432
Female-summer	0.038	< 0.001*	< 0.001*	0.187	0.001*	0.027
SIERRA NEVADA	1983	1984	1985	1986	1987	
Male-winter			0.166			
Male-summer			0.268			
Female-winter	0.246	0.294	0.607	0.659	0.785	
Female-summer		0.222	0.407	0.188	0.724	

Notes: * P-value < 0.05, indicating a significant difference.
P-values associated with one-way ANOVAs comparing river density (m/m^2) between exclusive and shared portions of mountain lion home ranges for each study area (Diablo 1984–1989, Santa Ana 1987–1992, Sierra Nevada 1983–1987), based on sex (male-female) and seasonal (winter-summer) differences.

Table 14.4. Significance of differences in number of habitat types between shared and exclusive home ranges

DIABLO RANGE	1984	1985	1986	1987	1988	1989
Male-winter	0.123	0.003*	0.003*	0.692	0.195	
Male-summer	0.82	0.136	0.03	0.311	0.502	
Female-winter	0.366	0.009	0.335	0.122	0.007*	
Female-summer	0.001*	0.035	0.514	0.083	0.029	< 0.001*
SANTA ANA	1987	1988	1989	1990	1991	1992
Male-winter				0.283	0.194	0.031
Male-summer				0.562	0.449	0.003*
Female-winter	0.02	< 0.001*	0.175	0.351	< 0.001*	0.075
Female-summer	0.044	< 0.001*	< 0.001*	< 0.001*	0.01	0.705
SIERRA NEVADA	1983	1984	1985	1986	1987	
Male-winter			0.172			
Male-summer			0.495	0.149		
Female-winter	0.711	0.394	0.645	0.423	0.326	
Female-summer		0.336	0.047	0.186	0.031	

Notes: *P-value < 0.05, indicating a significant difference.
P-values associated with one-way ANOVAs comparing the number of habitat types between exclusive and shared portions of mountain lion home ranges for each study area (Diablo 1984–1989, Santa Ana 1987–1992, Sierra Nevada 1983–1987), based on sex (male-female) and seasonal (winter-summer) differences.

DISCUSSION

Overall HR size for female mountain lions appears closely linked to food resources. During summer, females in the Sierra Nevada had larger HR sizes than those in both of the Coastal Range ecosystems. This pattern likely coincides with migratory patterns and HR expansions of black-tailed deer in the Sierra Nevada (Neal 1990; Pierce et al. 1999), whereas HR reductions of deer during dry months in the Coastal Ranges. During winter, the reverse pattern is evident; female HRs in the Sierra Nevada were smaller than those in Santa Ana. This pattern likely corresponds to deer in the Sierra Nevada reducing their winter HRs when concentrated in lower Sierra Nevada elevations (Neal 1990; Pierce et al. 1999), whereas deer in Santa Ana were more widespread in winter because water is more plentiful (Beier and Barrett 1993; Padley 1990). Home range size for females in the Diablo Range was consistently smaller than the other areas. This could be due to higher deer densities in the Diablo Range (10.5 deer/km^2) compared with the Sierra Nevada (0.868 deer/km^2) or Santa Ana (1.9–5 deer/km^2). In addition, the Diablo Range is far less developed than the Santa Ana. Padley (1990) suggests that HRs in the Santa Ana may expand to accommodate for poor quality and heavily developed mountain lion habitat.

Greater differences between ecological and anthropogenic habitat features in exclusive vs. shared portions of mountain lion HRs were detected in the Coastal Ranges (i.e., Diablo Range and Santa Ana) than in the Sierra Nevada. In the Coastal Ranges, there were significant differences between exclusive vs. shared portions of HRs in regard to road density, river density, and number of habitat types, while in the Sierra Nevada, the only differences detected were in road density (between exclusive and shared portions of the HR) during 2 out of 12 seasons. Hence, ecological variables analyzed during this study do not seem to be as important in determining habitat use for mountain lions in the Sierra Nevada as they are in the Coastal Ranges. In the Sierra Nevada, mountain lions follow migratory deer from lower elevations in winter to higher elevations in summer. This migration necessitates the need for mountain lions to create separate HRs for each season (Pierce et al. 1999). Since food resources for mountain lions in the Sierra Nevada are so wide ranging, they may not maintain exclusive areas long enough for fine-scale habitat features to have as much importance.

Despite these differences, the one aspect of habitat use for which no significant differences exist among studies is HR overlap: female and male mountain lions shared approximately one-fourth of their HRs with conspecifics (a little less for females in the Diablo Range). Home range over-lap in mountain lions is likely influenced by factors other than how variable ecological resources are in time and space. Spatial patterns in felids are often dependent on social factors, such as opportunities to mate (Sandell 1989), and therefore may be constant across different habitat types, and even in highly disturbed areas. The consistency of this sociospatial pattern likely indicates that population density is partially limited by territoriality—while HRs may shrink or expand in relation to prey distributions, for example, the proportion of the HR that is exclusive remains constant.

Implications for Sustainable Management of Mountain Lions

Our results indicate that there are significant differences in habitat use be-tween mountain lion populations in California. Conservation strategies based on fine-scale spatial and temporal knowledge of local mountain lion populations will best promote coexistence between people and lions in sub-urban communities (Carter et al. 2012). Also within suburban communities, the protection of appropriate corridors for dispersal (Santa Ana, Diablo) and migration (Sierra Nevada) of mountain lions would connect impor-tant habitats that in some areas, such as the Sierra Nevada, are occupied during different seasons (Neal 1990; Beier 1993; Pierce et al. 1999). In more developed areas, such as the Santa Ana, conservation plans should account for the possibility that mountain lions may require larger home ranges to meet their needs in marginal habitats, and that mountain lion population density may be reduced within these habitats. Consequently, mountain lion populations in highly developed areas may be at greater risk for inbreeding depression if dispersal between subpopulations is low, as is the case in the Santa Monica Mountains of California (Riley et al. 2014). If isolated habitat fragments are not adequate in size, it may be necessary to consider options such as translocation for individuals in these areas (Beier 1993).

Road density has the potential to curtail dispersal and mating opportuni-ties by restricting movement for many large mammals. Therefore, corridors placed appropriately across (under or over) major roadways are essential to long-term population viability of populations near urban areas (Beier 1993).

Working/agricultural landscapes should also be integrated into conser-vation plans for mountain lions and other carnivores (Hilty and Meren-lender 2004). Although marginal, these areas potentially could be utilized by species such as mountain lions in highly developed areas, and serve as important dispersal corridors. Marginal areas may be used in accordance to human presence. For example, in Kenya, lions (*Panthera leo*) used buffer areas primarily when people and livestock were absent. Seasonal changes into buffer areas by people/livestock corresponded with changes in lion occupancy out of buffer areas and into conservation areas (Schuette et al. 2013). Similarly, by changing their spatial and temporal activity, tigers (*Panthera tigris*) in Nepal have been shown to adapt and thrive in marginal areas dominated by humans (Carter et al. 2012).

In suburban landscapes it is vital to assess how fine-scale differences in land cover and development affect wildlife movement and to anticipate how future land use changes may increase habitat fragmentation. Increasing suburbanization may easily lead to genetic isolation, as has occurred in numerous species (e.g., American martens [*Martes americana*], Broquet et al. 2006; Australian squirrel gliders [*Petaurus norfolcensis*], Taylor et al. 2011; Japanese sika deer [*Cervus nippon*], Yuasa et al. 2007; mountain lions in California, Riley et al. 2014, and in Florida, Roelke, Martenson, and O'Brien 1993).

In carnivores, body size is positively correlated with sensitivity to fragmentation (Crooks 2002). If large carnivores are unable to disperse effectively in developed areas, populations are likely to diverge both genetically and behaviorally. As our results show, individuals of a species may vary greatly in habitat use. Therefore, data specific to local populations, dispersing juveniles, and migrating individuals should be used in designing corridors. Dispersing juveniles may be less risk averse than mature adults, and may even select against optimal habitat in order to avoid conspecifics, as is the case for African lions (Elliot et. al. 2014). For mountain lions, dispersers have been found to traverse areas of nonhabitat between subpopulations (Sweanor, Logan, and Hornocker 2000). Once again, temporal avoidance of humans can allow large carnivores to safely navigate developed landscapes. African lions, tigers, and mountain lions all shift their daily and seasonal activity patterns to increase utilization of developed areas at times when human activity is reduced (Carter et al. 2012; Schuette et al. 2013; Van Dyke 1984).

In summary, the best way to ensure the long-term sustainability of suburban carnivore populations is to provide corridors that allow dispersers to travel between semi-isolated subpopulations. Corridors for mountain lions do not need to be comprised of pristine habitat, and can incorporate moderately developed areas. Connecting the landscape for species such as mountain lions will require collaboration between researchers, managers, state and municipal officials, and local communities (Kays et al. 2015). Although suburban areas often support dense human settlements, Carter et al. (2012)

demonstrate that habitat management and lower exploitation may be more important to conserving species like tigers in Nepal than human density. Creating an urban/suburban landscape that carefully connects suitable and marginal habitats for species can promote dispersal and overall persistence for these populations. With cooperation from local communities, sustainable management of suburban wildlife populations is feasible as long as appropriate linkages are maintained between subpopulations.

ACKNOWLEDGMENTS

We wish to thank Dr. Paul Beier, Dr. Rick Hopkins, Mr. Don Neal, and Mr. Wayne Padley for their dedication to and comprehensive research on mountain lions in California. We also wish to thank Dr. Joshua Viers for his critical contributions to the manuscript.

REFERENCES CITED

Abay, Gidey Y., Hans Bauer, Kindeya Gebrihiwot, and Jozef Deckers. 2011. "Peri-urban Spotted Hyena (*Crocuta crocuta*) in Northern Ethiopia: Diet, Economic Impact, and Abundance." *European Journal of Wildlife Research* 57(4): 759–765.

Andrews, Kimberly M., J. Whitfield Gibbons, and Denim M. Jochimsen. 2008. "Ecological Effects of Roads on Amphibians and Reptiles: A Literature Review." In *Urban Herpetology*, edited by J. C. Mitchell, R. E. Jung Brown, and B. Bartholomew. Herpetological Conservation, vol. 3. Salt Lake City, UT: Society for the Study of Amphibians and Reptiles.

Beckmann, Jon P., and Carl W. Lackey. 2008. "Carnivores, Urban Landscapes, and Longitudinal Studies: A Case History of Black Bears." *Human-Wildlife Contacts* 2: 168–174.

Beier, Paul. 1993. "Determining Minimum Habitat Areas and Habitat Corridors for Cougars." *Conservation Biology* 7: 94–108.

Beier, Paul, and Reginald H. Barrett. 1993. "The Cougar in the Santa Ana Mountain Range, California." Final Report: Orange County Cooperative Mountain Lion Study. Berkeley: University of California, Berkeley.

Belden, Robert C., and James W. McCown. 1996. "Florida Panther Reintroduction Feasibility Study." Final Report: Bureau of Wildlife Research, Division of Wildlife, Florida Game and Fresh Water Fish Commission. Study Number: 7507.

Beschta, Robert L., and William J. Ripple. 2009. "Large Predators and Trophic Cascades in Terrestrial Ecosystems of the Western United States." *Biological Conservation* 142(11): 2401–2414.

Broquet, Thomas, Nicolas Ray, Eric Petit, John M. Fryxell, and Francoise Burel. 2006. "Genetic Isolation by Distance and Landscape Connectivity in the American Marten (*Martes americana*)." *Landscape Ecology* 21: 877–889.

Burdett, Christopher L., Kevin R. Crooks, David M. Theobald, Kenneth R. Wilson, Erin E. Boydston, Lisa M. Lyren, Robert N. Fisher, T. Vickers, Morrison Winston, Scott A.

Morrison, and Walter M. Boyce. 2010. "Interfacing Models of Wildlife Habitat and Human Development to Predict the Future Distribution of Puma Habitat." *Ecosphere* 1: 1–21.

CALHOME. 1994. MS-DOS Version 1.0. Forestry Sciences Lab, Fresno, CA.

Carter, Neil H., Binoj K. Shrestha, Jhamak B. Karki, N.M.B. Pradhan, and J. Liu. 2012. "Coexistence Between Wildlife and Humans at Fine Spatial Scales." *Proceedings of the National Academy of Sciences of the United States of America* 109(38): 15360–15365.

Cirovic, Dusko, Aleksandra Penezic, and Miha Krofel. 2016. "Jackals as Cleaners: Ecosystem Services Provided by a Mesocarnivore in Human-dominated Landscapes." *Biological Conservation* 199: 51–55.

Crooks, Kevin R. 2002. "Relative Sensitivities of Mammalian Carnivores to Habitat Fragmentation." *Conservation Biology* 16(2): 488–502.

Cypher, Brian L. 2010. "Kit Foxes (*Vulpes macrotis*)." In *Urban Carnivores: Ecology, Conflict, and Conservation,* edited by Seth P. D. Riley, Stanley D. Gehrt, and Brian L. Cypher, 223–232. Baltimore, MD: The Johns Hopkins University Press.

Elliot, Nicholas B., Samuel A. Cushman, David W. Macdonald, and Andrew M. Loveridge. 2014. "The Devil Is in the Dispersers: Predictions of Landscape Connectivity Change with Demography." *Journal of Applied Ecology* 51(5): 1169–1178.

Estes, James A., John Terborgh, Justin S. Brashares, Mary E. Power, Joel Berger, William J. Bond, Stephen R. Carpenter, Timothy E. Essington, Robert D. Holt, Jeremy B. C. Jackson, et al. 2011. "Trophic Downgrading of Planet Earth." *Science* 333(6040): 301–306.

Fernandez-Juricic, Esteban, and Jukka Jokimaki. 2001. "A Habitat Island Approach to Conserving Birds in Urban Landscapes: Case Studies from Southern and Northern Europe." *Biodiversity and Conservation* 10: 2023–2043.

Hilty, Jodi A., and Adina M. Merenlender. 2004. "Use of Riparian Corridors and Vineyards by Mammalian Predators in Northern California." *Conservation Biology* 18: 126–135.

Hopkins, Rick A. 1989. "Ecology of the Puma in the Diablo Range, California." PhD dissertation, University of California, Berkeley.

Hopkins, Rick A., M. J. Kutilek, and G. L. Shreve. 1986. "Density and Home Range Characteristics of Mountain Lions in the Diablo Range of California." In *Cats of the World: Biology, Conservation and Management,* edited by S. D. Miller and D. D. Everett, 223–235. Kingsville, TX: Caesare Kleberg Wildlife Research Institute.

Kays, Roland, Margaret C. Crofoot, Walter Jetz, and Martin Wikelski. 2015. "Terrestrial Animal Tracking as an Eye on Life and Planet." *Science* 438(6240): aaa2478.

Kertson, Brian N., Rocky D. Spencer, John M. Marzluff, Jeff Hepinstall-Cymerman, and Christian E. Grue. 2011. "Cougar Space Use and Movements in the Wildland-Urban Landscape of Western Washington." *Ecological Applications* 21(8): 2866–2881.

Levi, Taal, A. M. Kilpatrick, M. Mangel, and C. C. Wilmers. 2012. "Deer, Predators, and the Emergence of Lyme Disease." *Proceedings of the National Academy of Sciences* 109: 10942–10947.

NCSS. 1996. Number Cruncher Statistical System (version 6.0.21). Windows. Kaysville, UT.

Neal, Donald L. 1990. "The Effect of Predation on Deer in the Central Sierra Nevada." In *Predator Management in North Coastal California: Proceedings of a Workshop Held in Ukiah and Hopland, California, March 10–11, 1990,* edited by Gregory A. Giusti, Robert

M. Timm, and Robert H. Schmidt, 53–61. Hopland: University of California: Hopland Field Station Publications.

Padley, Wayne D. 1990. "Home Ranges and Social Interactions of Mountain Lions (*Felis concolor*) in the Santa Ana Mountains, California." Master's thesis, California State Polytechnic University, Pomona.

Pierce, Becky M., Vernon C. Bleich, John D. Wehausen, and R. Terry Bowyer. 1999. "Migratory Patterns of Mountain Lions: Implications for Social Regulation and Conservation." *Journal of Mammalogy* 80: 986–992.

Purvis, Andy, John L. Gittleman, Guy Cowlishaw, and Georgina M. Mace. 2000. "Predicting Extinction Risk in Declining Species." *Proceedings of the Royal Society B* 267(1456): 1947–1952.

Riley, Seth P. D., Stanley D. Gehrt, and Brian L. Cypher. 2010. "Urban Carnivores: Final Perspectives and Future Directions." In *Urban Carnivores: Ecology, Conflict, and Conservation*, edited by Seth P. D. Riley, Stanley D. Gehrt, and Brian L. Cypher, 223–232. Baltimore, MD: The Johns Hopkins University Press.

Riley, Seth P. D., Laurel E. K. Serieys, John P. Pollinger, Jeffrey A. Sikich, Lisa Dalbeck, Robert K. Wayne, and Holly B. Ernest. 2014. "Individual Behaviors Dominate the Dynamics of an Urban Mountain Lion Population Isolated by Roads." *Current Biology* 24(17): 1989–1994.

Ripple, William J., James A. Estes, Robert L. Beschta, Christopher C. Wilmers, Euan G. Ritchie, Mark Hebblewhite, Joel Berger, Bodil Elmhagen, Mike Letnic, Michael P. Nelson, et al. 2014. "Status and Ecological Effects of the World's Largest Carnivores." *Science* 343(6167): 12414841–124148411.

Roelke, Melody E., Janice S. Martenson, and Stephen J. O'Brien. 1993. "The Consequences of Demographic Reduction and Genetic Depletion in the Endangered Florida Panther." *Current Biology* 6(1): 340–350.

Sandell, Mikael. 1989. "The Mating Tactics and Spacing Patterns of Solitary Carnivores." In *Carnivore Behavior, Ecology, and Evolution*, edited by John L. Gittleman, 164–182. Ithaca, NY: Comstock Publishing Associates, Cornell University Press.

Schipper, Jan, Janice S. Chanson, Frederica Chiozza, Neil A. Cox, Michael Hoffmann, Vineet Katariya, John Lamoreux, Ana S. L. Rodrigues, Simon N. Stuart, Helen J. Temple, et al. 2008. "The Status of the World's Land and Marine Mammals: Diversity, Threat, and Knowledge." *Science* 322: 225–230.

Schuette, Paul, Aaron P. Wagner, Meredith E. Wagner, and Scott Creel. 2013. "Occupancy Patterns and Niche Partitioning Within a Diverse Carnivore Community Exposed to Anthropogenic Pressures." *Biological Conservation* 158: 301–312.

Smallwood, K. Shawn. 1994. "Trends in Mountain Lion Populations." *Southwestern Naturalist* 39: 67–72.

Spreadbury, Brian R., R. R. Kevin Musil, Jim Musil, Chris Kaisner, and John Kovak. 1996. "Cougar Population Characteristics in Southeastern British Columbia." *Journal of Wildlife Management* 60: 962–969.

Sweanor, Linda L., Kenneth A. Logan, and Maurice G. Hornocker. 2000. "Cougar Dispersal Patterns, Metapopulation Dynamics, and Conservation." *Conservation Biology* 14(3): 798–808.

Taylor, Andrea C., Faith M. Walker, Ross L. Goldingay, Tina Ball, and Rodney van der

Rhee. 2011. "Degree of Landscape Fragmentation Influences Genetic Isolation Among Populations of a Gliding Mammal." *PLoS ONE* 6(10): e26651. https://doi.org/10.1371/journal.pone.0026651.

Terborgh, John, Lawrence Lopez, Percy V. Nunez, Madhu Rao, Ghazala Shahabuddin, Gabriela Orihuela, Mailen Riveros, Rafael Ascanio, Greg H. Adler, Thomas D. Lambert, and Luis Balbas. 2001. "Ecological Meltdown in Predator-Free Forest Fragments." *Science* 294: 1923–1926.

Thorne, James H., Dick Cameron, and James F. Quinn. 2006. "A Conservation Design for the Central Coast of California and the Evaluation of Mountain Lion as an Umbrella Species." *Natural Areas Journal* 26: 137–148.

Van Dyke, Fred G. 1984. "A Western Study of Cougar Track Surveys and Environmental Disturbances Affecting Cougars Related to the Status of the Eastern Cougar *Felis concolor couguar*." PhD dissertation, State University of New York College of Environmental Science and Forestry, Syracuse.

Van Dyke, Fred G., Ranier H. Brocke, Harley G. Shaw, Bruce B. Ackerman, Thomas P. Hemker, and Frederick G. Lindzey. 1986. "Reactions of Mountain Lions to Logging and Human Activity." *Journal of Wildlife Management* 50: 95–102.

Vierling, Kerri T. 2000. "Source and Sink Habitats of Red-Winged Blackbirds in a Rural/Suburban Landscape." *Ecological Applications* 10: 1211–1218.

Yuasa, T., Junco Nagata, S. Hamasaki, S. Tsuruga, and K. Furubayashi. 2007. "The Impact of Habitat Fragmentation on Genetic Structure of the Japanese Sika Deer (*Cervus nippon*) in Southern Kantoh, Revealed by Mitochondrial D-loop Sequences." *Ecological Research* 22(1): 97–106.

15

Suburban Unsustainability or the Burden of Fixed Infrastructure in New Orleans, Louisiana

CRAIG E. COLTEN

There exists a powerful conundrum along the United States (U.S.) Gulf Coast. Vulnerable rural populations who subsist on locally abundant natural resources along the coast have endured hurricanes, river floods, oil spills, and human-made disasters, and they persist in place. This reveals a certain degree of resilience—abilities to flex in the face of extreme events, adapt, rebuild, and recover. They have built stilt houses and many are essentially self-insured. That is they did not invest in elaborate houses with expensive furnishings and belongings. Many simply could not afford all the trappings of affluence. In some respects, modest circumstances enabled a more sustainable lifestyle.

Yet urban residents in the region have not followed the lead of their rural counterparts. Postwar urbanization has sprawled out of the safer settings of older urban cores. Developers erected thousands of slab-on-grade ranch houses across floodplains, toward the coast, and in greater densities. Urban coastal populations have risen dramatically in recent decades (NOAA 2013). With rising sea levels, particularly in coastal Louisiana which has a subsiding shoreline and also the risks connected with low-lying property, populations and infrastructure from the Florida Keys to Brownsville, Texas face heightened risks. Floods in Louisiana (2016), Texas (2017), and Florida (2017) have highlighted the extensive costs and personal disruptions faced by under-insured homeowners living on flood-prone real estate with inadequate flood mitigation infrastructure.

Sustainability was not a factor for securing post–World War II VA loans, nor was it a planning consideration during the expansive suburbanization of the 1960s and 1970s. Public officials frequently granted developers rights to expand their urban tax base and turned to federal authorities for flood protection after the fact. To a degree, the National Flood Insurance Program standards sought to minimize risk as that program ramped up in the 1980s and 1990s, but sustainability was not part of the typical flood mitigation deliberations. Consequently, most of the postwar suburban infrastructure shares a path dependency toward unsustainability. This chapter considers the processes that contributed to this regional pattern of over-development by focusing on the New Orleans metropolitan area which faces tropical cyclones and intense precipitation that can produce local flooding and widespread damage while it is also undergoing regional subsidence. A long-term perspective offers vital insights into the local culture of flood management and how past actions perpetuate risk and policies that foreground development ahead of risk reduction and public safety.

Suburban Sustainability

In 1999 the National Research Council published an overview of what it termed the sustainability transition and asked the question: can the people living on earth over the next half-century meet their needs while nurturing and restoring the planet's life support systems? The report claimed that the transition was foregrounding sustainability in public and political discussions and was fundamentally a "reconciliation of society's developmental goals with the planet's environmental limits over the long term" (NRC 1999, 1–2). It posited that human decisions could guide this sustainability transition which it described as an ongoing, adaptive process and not a static goal. This report offered its most fundamental goal as meeting: "the needs of a much larger but stabilizing human population, to sustain the life support systems of the planet, and to substantially reduce hunger and poverty" (NRC 1999, 4). At the core of its approach was the recognition of the need to conduct research that integrates local and global perspectives in a place-based understanding of human-environment interactions.

In terms of cities, there exists competing views about urban sustainability, but both acknowledge the challenges of suburbs. Geographer William Meyer points out that cities concentrate people and pollution; they create urban heat islands; technological hazards beset them; they enable the rapid spread of infectious disease; and they consume huge quantities of natural resources. Yet, he concludes that cities "lessen pressure on ecological systems by confining it in space, they slow population growth, and they make the consumption of natural resources more sparing and efficient" (Meyer 2013, 146). Benton-Short, Keeley, and Rowland (2017) in a similar vein contend that cities around the globe have been at the forefront of developing sustainability plans. Environmental scientists John Day and Charles Hall argue that the notion that cities offer environmental efficiency, and therefore offer a hope for sustainable settlements, is a myth (Day and Hall 2016, 26). They make the case that cities were built on cheap energy and that in the post-peak oil era their larger carbon footprints offset any savings through their concentration of people and infrastructure (Day and Hall 2016, 25–36). They claim that cities, with extensive suburbs, and their surrounding urban regions, can never be self-sustaining or even sustainable. None of these investigators makes the pretense of offering a hope for true urban self-sufficiency, nonetheless one argument is that by charting a path toward a more sustainable future, cities can produce notable accomplishments because they house the most people.

How should we approach suburbs as part of this discussion of urban sustainability? Suburbs, as part of the larger urban landscape, have earned a reputation as the most troublesome and unsustainable landscapes. As a whole, they are land extensive and wasteful resource sinks. Nineteenth century suburbs, however, escape much of this criticism owing to their more compact footprint. The sprawling post–World War II suburbs are the chief targets of criticism, and there has arisen a literature devoted to addressing this condition. Planners and geographers have launched into discussions about how to retrofit these unsustainable landscapes into viable spaces that will not impede their use by future generations (Talen 2011). The related topic of "green infrastructure" has neglected risk reduction structures which must be part of the sustainability discussion (Short 2017). Also of importance are the lasting effects of antiquated infrastructure that is neither easy to remove or upgrade. Structural flood protection offers a poignant example of the path dependence that has contributed to past suburban growth (Melosi 2000) and future suburban risks (Colten 2009).

There has been ample discussion about the concentration of urban populations in coastal regions that face increasing risk from sea level rise (NOAA 2013). And there has been careful analysis of risk to cities and their infrastructure (O'Neill and Abs 2016; Colten 2009; Mitchell 1999). One means to gain perspective on the challenges of future adaptation to rising sea levels is to consider the long-term adaptations and risk reduction efforts carried out in a region that has contributed to its own human-induced sea level rise—through land subsidence and coastal erosion. The New Orleans metropolitan area, including its postwar suburbs, has experienced serious land subsidence within its river and hurricane protection levees. Some areas have subsided as much as 15 feet and are now well below sea level. Levees and pumps keep these areas relatively dry most of the time. Investment in these flood protection structures, and plans for extensive coastal restoration facilities to further mitigate risk, enable urban life to continue in this coastal location. Do such investments provide a sustainable fix to suburban areas facing sea level rise?

PROTECTING A CITY AND ITS SUBURBS FROM FLOODS

From its inception as a colonial settlement, New Orleans has faced the dual threats of river flooding and hurricane-induced storm surge. To contend with the almost annual risk of the Mississippi River overtopping its banks and flowing through the city's streets, French colonial officials erected modest levees along the urban waterfront, and required rural land owners to fortify their own riverfront properties to create a somewhat continuous line of protection for the lower river region. Colonial-era levees were inconsistent in strength and height and offered limited protection. Frequent breaches prompted constant repair, which did not offer a sustainable solution. Indeed, as laborers added height to the levees and they rose to a more formidable stature, confinement prompted the river to raise its bed, which exacerbated risk, and required perpetual raising of the levee heights (Colten 2005, 16–46). A more consistent levee system was the object of the Mississippi River Commission, formed in 1879 (Camillo and Pearcy 2004). But its massive efforts to fortify the levee system were inadequate to contain the record flood of 1927. This tragedy triggered a reformulation of the flood protection system from a "levees only" to a "levees and outlets" approach. Up until 1928, the federal system had closed off natural outlets and limited levees to the main channel of the Mississippi River. This option disconnected the river from several distributaries which had functioned as "safety valves" that could carry excess flow during high stages. The levees and outlets approach introduced a pair of human-made outlets that the Corps of Engineers could open during exceptional floods. As the levees themselves, these structures demanded massive investments and ongoing maintenance and repair. During the 1973 flood, the Old River Control Structure, one of the engineering works that was vital to the operation of the larger of the two outlets, almost failed (Barnett 2017). This near disaster and the subsequent costly repairs underscored the unsustainable condition of this expensive flood protection system.

Hurricanes caused extensive damage in New Orleans from the 1700s on, but their irregular arrival and inconsistent impacts deterred construction of mitigation structures (Rohland 2019. A hurricane in 1915 washed over the wetlands that buffered the city from Lake Pontchartrain and flooded low-lying neighborhoods, and prompted New Orleans officials to adapt levees to hurricane flood protection (Figure 15.1). Over the next two decades, they secured funding and built a seawall along the lakefront. It included both a sinuous stepped concrete wall that rose 9.5 feet above the lake level and an extensive swath of "made land" behind the barrier. This newly created real estate was marketed as suburban lots to underwrite the cost of the protective structure. Thus, flood protection infrastructure gave birth to new suburbs (Colten 2012).

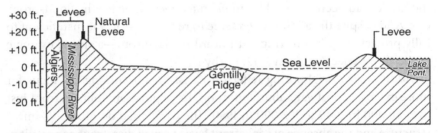

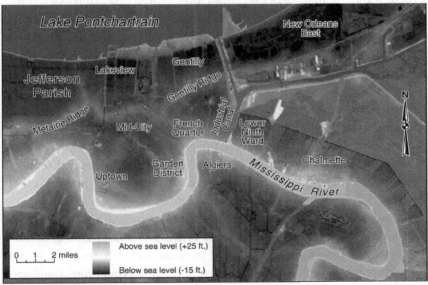

Figure 15.1. New Orleans topography and suburbs. Cartography by Mary Lee Eggart and Dewitt Braud.

Completed before the next major storm in 1947, the seawall reduced flooding to a modest degree, but local planners complained that the most extensive damage was to new houses that were not built like the more traditional structures in New Orleans (Orleans Levee District 1950). In recognition of its frequent flood risks in low-lying neighborhoods and often saturated soils, builders had commonly erected houses elevated on piers. Some houses stood five to six feet above grade, while even the more ubiquitous shotgun and double-shotgun houses were elevated about two feet off the ground. Even the lower dwellings were flood proofed against the most common inundations. The houses that suffered damage in 1947 conformed to national building practices that had become the new standard and had concrete slabs directly on the ground and stood, maybe, six inches above the soil. A tradition of building elevated houses, that was well adapted to the locale, had been replaced by an ill-suited technique for the flood-prone city. And despite the addition of a levee to reduce risk, it was insufficient to fully protect suburbs growing out toward the lakefront—a classic example of the "levee effect" (Colten 2012; Burby 2006).

Additionally, neighboring Jefferson Parish was becoming the focus for more extensive postwar suburbanization. Its population doubled from just over 50,000 in 1940 to 103,000 in 1950. The parish had outgrown its infrastructure and the absence of a lakefront levee proved disastrous as a result of the 1947 hurricane. Several thousand homes sustained substantial damage when storm surge overtopped the 3 foot high berm supporting a lakefront highway. Local officials appealed to the Corps of Engineers to build protection comparable to the seawall in New Orleans (U.S. Congress 1950). The project received congressional support and moved forward. By 1956, some authorities, however, were questioning the adequacy of this new levee. A local official testified to the Corps that thousands of new homes now stood behind the levees and warned that if the protection system failed, residents would lose untold amounts of property (U.S. Army Corps of Engineers 1956). Postwar suburbs depended on levees that some officials lacked confidence in only a few years after their completion.

As suburbs spread across east bank Jefferson Parish during the next few decades, the population doubled again between 1950 and 1960 (208,000). Mostly white suburbanites purchased thousands of slab-on-grade houses behind the newly built levees not realizing that the flood protection barriers contributed to a new flood hazard: subsidence. Since levees eliminated natural drainage from what had been marsh lands, the parish had to build extensive drainage canals to carry water and massive pumping stations to lift water into the lake. Leveeing and draining the former wetlands produced wholesale subsidence in the new suburbs. These environmental changes caused houses built on peaty soils to settle which damaged both properties and infrastructure. In response, local governments required the use of wooden friction piles driven into the soils before construction in order to stabilize foundations. This technique worked well, but subsidence continued and there was a spate of explosions as sinking land ripped gas lines out of stationary house foundations (Snowden, Ward, and Studlick 1980). New policies that addressed gas line connections eventually abated this situation, and reflected adaptation to one of the many challenges of building suburbs on subsiding, peaty soils.

When Hurricane Betsy slammed into southeast Louisiana in 1965, its winds of 120 miles per hour caused wind damage, but inadequate hurricane protection levees or gaps in the existing levees allowed water to enter the city that caused the greatest harm to properties. Some 15,000 home sustained serious damage and floodwaters covered over 40 percent of New Orleans. The Corps had submitted a plan for an expanded levee system before the storm, and congress hastily approved the plan in the wake of the dramatic storm (Figure 15.2). The only way to justify expenditures on an extensive and expensive levee system to surround much of the undeveloped wetlands that constituted eastern New Orleans and adjacent parishes was to tabulate the values of anticipated suburban development. Estimates placed damages that would be prevented annually over the life of the project to be $47 million, and this figure secured a favorable cost-benefit analysis (U.S. Army Corps of Engineers 1962; U.S. Army Corps of Engineers 1966).

Work began quickly on levees that surrounded largely uninhabited areas of eastern New Orleans, and development followed. During the next several years, the city approved a spate of new subdivisions directly on the footprint of areas that had been flooded in 1965. Between 1965 and the mid 1980s, when residential construction stalled due to the oil bust, contractors erected some 22,000 new slab-on-grade homes in the booming suburbs of New Orleans East (Burby 2006). Federal funds offered enhanced flood protection and local developers swarmed to take advantage of new opportunities. Subsidence in eastern New Orleans caused foundation failures and other infrastructure problems. These unintended consequences dampened enthusiasm among property buyers, and real estate prices fell especially with the advent of the oil bust in the 1980s. In some respects, the lower prices and the eagerness of realtors to move properties created a new opportunity for middle-class African Americans seeking the suburban life. New Orleans

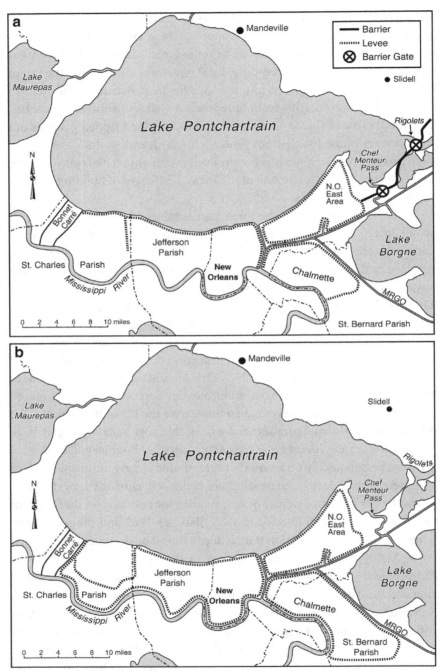

Figure 15.2. New Orleans hurricane protection plans: (*a*) barrier plan, (*b*) high-level plan. Cartography by Mary Lee Eggart.

East became a predominately African American suburb within a ring of levees, bounded by Lake Pontchartrain on the north and the MR-GO navigational waterway to the south that some writers have referred to as a hurricane superhighway (Campanella 2007; Shallat 2000).

West bank Jefferson Parish offers numerous contrasts to its east bank counterpart. Upstream from the parish seat of Gretna, industries hug the riverbank and occupy the highest ground on the natural levees. There were working class neighborhoods near the refineries and other manufacturing facilities and these became largely African American during the postwar years. The west bank's population leaped from 32,000 in 1940 to 180,000 in 1980. As elsewhere, suburban expansion in this portion of the parish required drainage and other structural protections to extend the habitable footprint into the neighboring wetlands.

As the Corps planned its hurricane protection for New Orleans proper and east bank Jefferson Parish, it also deliberated a levee to guard the west bank area from storm surge that could invade the suburbs from the wetlands to the south. The Corps presented several levee options to local authorities in 1972, but they rejected them expressing dissatisfaction with the small territory to be protected—and levees in this setting define the limits of development and taxable real estate. After a protracted process involving coordination with other federal projects, federal engineers proposed a second levee alignment in 1979. Once again local officials declined the offer and approved taking on the full task of financing and building the levees. They hoped their actions would both accelerate the process and expand the taxable territory. Their plans ran aground when the parish submitted its environmental impact statement to the Corps in 1984. The federal engineers declined the parish's application, but offered a compromise that would protect nearly 2,000 acres of wetland that had been targeted for development in the parish's plan. With the loss of a sizable, taxable territory, the parish decided to turn everything back over the Corps in 1985. Some 20 years after post–Hurricane Betsy planning for protecting the west bank began, no real progress had been made. A modest fall hurricane (Juan) that year threatened the west bank and intensified efforts to build protection for this suburban section of the city (Colten 2009, 94–98).

In addition to hurricanes, heavy downpours could threaten the expanding west bank suburbs. A massive storm in 1978 exposed the risk of pushing suburbs into former swamps. The combination of construction and pavement, plus a drainage system that was inadequate to the task of moving out a massive rainfall resulted in damage to some 30,000 homes ("Area Flood

Price Tag $240 Million" 1978; "Must Limit West Bank" 1978; "Floods Not the Only Problem in Marrero Disaster Area" 1978).

Racial politics played a part in the subsequent response to the recognized flood risk. Predominately white, east bank Jefferson Parish sustained little damage from the 1978 storm and voted overwhelmingly against a tax hike to fund drainage improvements. Two massive rain storms in April 1980 further emphasized the inadequacies of the drainage infrastructure when they caused damage exceeding $150 million in west bank neighborhoods. A large number of home owners had flood insurance, but FEMA had reached its limit with repetitive flood claims and filed suit against the parish claiming its land use and construction practices were not in compliance with NFIP standards. After an extended legal battle, FEMA and the parish reached a settlement that called for the local government to pay the federal agency $1 million and to develop a flood reduction plan that would upgrade flood-proof building practices and prohibit construction in the 100-year floodplain ("Parishes Get Break in U.S. Flood Case" 1985; "Court Overturns Liability" 1985; U.S. v. Jefferson Parish et al. 1987).

Suburban development slowed throughout the metro area during the economic turmoil of the 1980s, but rainfall did not. The parish continued to rely on structural approaches—levees and drainage systems, but was unable to keep pace with the runoff from copious precipitation events. A particularly massive storm in 1995 caused some $545 million damage across the parish. This prompted the officials to enter into a regional drainage program administered by the Corps of Engineers. The Southeast Louisiana Flood Control Project maintained primary reliance on structural means and dedicated some $170 million to improvements to Jefferson Parish, despite the NFIP emphasis on land use controls. Limited land for development prioritized canals, pumps, and levees to expand the footprint of suburban real estate. Huge rainfall delivered by tropical weather systems in 2001 and 2002 caused only minor street flooding in Jefferson Parish, suggesting these improvements were effective (Colten 2005, 159–160).

INFRASTRUCTURE AND AGILE ADAPTATION

Building a massive earthen levee system around the New Orleans metropolitan area comes with innumerable engineering challenges. One of the most obvious is the sheer weight of the levees themselves, heaped atop saturated marshy soils. These soils could not adequately bear the weight of suburban houses constructed on concrete slabs, let alone the greater weight of levees.

As in the case of Jefferson Parish, local interests complicated the task of engineering and building protective barriers, and these political conflicts protracted the process of completing the hurricane protection system. Once in place, it is hard to adapt suburban infrastructure to achieve a more sustainable landscape. Consequently, historical decisions and actions came back to haunt tens of thousands of residents of New Orleans and its suburbs when Hurricane Katrina moved unimaginable amounts of water toward, over, and through the hurricane protection system in 2005.

The initial post-Betsy hurricane protection system design called for new or enlarged levees to surround most of the urban and suburban territory of greater New Orleans (Figure 15.2). It also called for barriers that could be closed to prevent storm surge from entering Lake Pontchartrain from the Gulf of Mexico and overwhelming the flood susceptible lakefront neighborhoods. By building barriers, the levee heights along the lake would not have to be as high and consequently not as expensive. Central to the entire enterprise was a desire to defend the existing city, while allowing further suburban expansion. The Corps' cost-benefit analysis, as in previous projects reached a favorable ratio only by projecting the benefits of future flood damages that would be prevented by the project. For the entire project, the Corps projected that the levees would enable $65 million dollars of combined development and flood prevention (U.S. Army Corps of Engineers 1967).

Congress gave the green light to funding and the Corps and local partners set to work enlarging the levee perimeter. Local challenges emerged and forced delays and modifications to the plan. Some organizations opposed the barrier plan and argued that draining wetlands in eastern New Orleans would be a public subsidy for private developers and unnecessarily destroy extensive wetlands (U.S. Army Corps of Engineers 1975). A consortium of opponents filed a lawsuit which challenged the Corps' 1974 environmental impacts statement for the "barrier plan." Opponents charged that periodic closure of the barriers would damage the lake's ecology, harm commercial fisheries, and divert surge to communities on the lake's north shore. The core legal argument was that the Corps' EIS was inadequate since it did not take these concerns into account. A judge issued an injunction on the barrier portion of the levee system in 1977. This judicial action prompted the Corps to re-evaluate its plan, and ultimately it decided to drop the barrier option and instead proceed with an alternate "high-level" plan. This plan simply required higher levees along the lake front and no barrier. By the mid 1980s, the Corps had shifted its construction efforts to the high-level plan

(Save Our Wetlands Inc. et al. v. Early J. Rush III et al. 1977; Save Our Wet-lands Inc. et al. v. Early J. Rush III et al. 1978; U.S. Army Corps of Engineers 1984).

With the high-level plan (Figure 15.2) came deviations from initial de-signs. Army engineers recommended substituting gates at the mouth of the several outfall canals that directed runoff from New Orleans into Lake Pon-tchartrain. These gates could be closed during a hurricane and would pre-vent surge from entering the city from the lake. Yet, if closed the gates would paralyze the drainage system and cause flooding in the city by preventing removal of runoff. The local levee district and drainage authorities strongly opposed the gates. By 1990, the Corps dropped the gates from its designs. Without these closures, levees along the several outlet canals for the city's drainage system had to be raised, but it was impossible to build earthen levees with their broad footprint that would consume the property of resi-dences backing up to the canals. Suburban properties forced a recasting of the design. Consequently, the Corps installed floodwalls atop the exist-ing levees. These additions were a combination of sheet steel pilings driven into the soil through the existing levees and reinforced concrete walls set atop the levees which added the necessary height (Colten 2009, 78). These flood walls failed during Katrina when surge forced water into the canals and simply pushed the floodwalls aside, unleashing tragic flooding in the middle-class suburbs of the Lakeview area.

Levees surrounding the eastern New Orleans suburbs and the rapidly growing suburbs in St. Bernard Parish required additional height as well and presented different challenges. In order to build the earthen barriers to the Corps' design heights required several phases of construction. The satu-rated wetland soils could not support the bulk of the levees. To cope with the poor geologic foundation, the Corps built the levees to the prescribed design height, allowed them to settle for several years, and then "lifted" them by adding more soil atop the existing structures. This procedure had to be repeated multiple times over the course of several years (Colten 2009, 74). All in all, carrying out multiple lifts delayed completion of the project. Surge that overtopped levee segments in the eastern suburbs during Katrina occurred in areas where levees had settled below design height and were in between "lifts" (U.S. Army Corps of Engineers 2009).

Overall, the levees and their associated engineering works were a piece-meal system. Built in stages, under different design concepts, and modi-fied along the way. Maintenance was largely the responsibility of local levee boards, but some held design flaws that contributed to their failure in 2005

(Seed et al. 2005; Van Heerden et al. 2006; U.S. Army Corp of Engineers 2009). Inundation of 80 percent of the city followed levee failures in 2005. Repairs and modifications have provided some risk reduction in the aftermath of Katrina, but the structural system still has design limits and suburbs continue to face serious threats from massive storms. The primacy of development sometimes overshadowed safety and risk reduction. Although not couched in these terms, an expanding urban footprint diminished the sustainability of the system.

PUMPS AND DRAINAGE IN POST-KATRINA NEW ORLEANS

A recent episode of urban flooding in August 2017 underscored the role of old infrastructure on suburban properties. During the peak of the hurricane season on August 5, a summer thunderstorm dumped about 10 inches of rainfall on New Orleans, five inches in 3 hours in some locations. The Sewerage and Water Board has a massive, yet ancient system of canals and pumps designed to remove about one inch of precipitation in the first hour and about 0.5 inches per hour after that. The capacity of the drains was improved in the 1990s with the Corps' SELA project and further improvements were under way. Yet, the storm in August produced serious flooding, beyond the chronic, short-term street closures that are common. It took over 14 hours to drain the accumulated rainfall, and in the meantime water damaged over 200 houses and additional cars, and streets remained under water for extended periods of time. The Lakeview and Gentilly neighborhoods, areas that suffered major damage due to Katrina, endured serious flooding again. The drainage system was apparently not up to the task. Many of the drainage canals were clogged with debris due to inadequate maintenance and pumps in some of the worst-hit areas were only operating a half capacity which greatly reduced the movement water. In all, eight pumps were inoperative at the time of the deluge. The reduced pumping capacity was exacerbated a few days later when one of the turbines for the drainage system's power supply caught fire and went off-line ("Here's Why New Orleans Flooding Was so Bad, and Why People Are Being Fired over It" 2017; "Everything You Need to Know About New Orleans Pumps, Plant That Helps Power Them" 2017).

Many of the pumps are over 100 years old and the power system relies partially on early 20th century 25-cycle power. The antiquated system is an amazing historical relic, but demands meticulous maintenance and care. Funding and staffing problems have compromised the drainage authority's ability to keep up with the heavy demands on the system. Additional rain events prompted rising public apprehension, particularly as Hurricane Harvey approached Houston that same month. As with the levees, this infrastructure is an imposing system that is more difficult and expensive to replace than to continue patching. Yet, it further illustrates the gargantuan challenges cities face to continue serving far-flung suburbs and the near impossibility of making them truly sustainable. If the 50 inches of Hurricane Harvey rainfall unleashed over Houston had fallen on New Orleans with its weakened drainage system, the consequences would have been catastrophic ("Here's Why New Orleans Flooding Was so Bad, and Why People Are Being Fired over It" 2017; "Everything You Need to Know About New Orleans Pumps, Plant That Helps Power Them" 2017). Despite an ambitious plan to learn to "live with water" in New Orleans (Waggoner and Ball Architects 2017; Louisiana Office of Community Development 2017), the infrastructure consisting of levees, pumps, and drainage canals, continue to prioritize suburban living without water.

Conclusions

Urban infrastructure is designed to last, but often not as long as it is called on to serve. Rising costs in replacing vital infrastructure has prompted most urban regions to repair and selectively update or modify existing facilities. This is particularly true in the New Orleans metropolitan area where structural hurricane flood protection that has been designed and installed since the 1930s inhibits city-wide adaptations. There is a considerable effort to transform land use activities within the levee system that will be more sustainable, but any adjustments within the levees still depends on the massive barrier and drainage systems.

Levees begat suburbs. Indeed, justification for funding the hurricane protection system depended on projected residential and commercial development beyond the World War II-era urban footprint. Risk in the urban region is compounded by subsidence and rising sea levels and both the levees and the pumping system are not easily adapted or retrofitted to the gradual changes in environmental conditions. Sustainability remains a distant goal.

The long-term lessons that might be learned from the New Orleans situation include several general concepts that can be considered in other places with different physical and human landscapes. Older suburbs, with greater concentrations of people and structures are less costly to protect with structural hazard mitigation systems. Designing suburbs—both in terms of architecture and land use—with local biophysical factors and hazard risks in mind can reduce mitigation and recovery costs. Structural mitigation tends to offer limited protection that can become obsolete with changing regional land use and environmental conditions. Mitigation planning at a regional level can offer a more unified result than piecemeal locality based plans. Thoughtful, advance planning is not always feasible, but should become a standard part of the conversation between developers and hazards managers.

References Cited

"Area Flood Price Tag $240 Million." 1978. *New Orleans Times-Picayune*, May 6, 1.

Barnett, James F. Jr. 2017. *Beyond Control: The Mississippi River's New Channel to the Gulf of Mexico*. Jackson: University Press of Mississippi.

Benton-Short, Lisa, Melissa Keeley, and Jennifer Rowland. 2017. "Green Infrastructure, Green Space, and Sustainable Urbanism." *Urban Geography* 38: doi.org/10.1080/02723 638.2017.1360105

Burby, Raymond. 2006. "Hurricane Katrina and the Paradoxes of Government Disaster Policy." *Annals of the Association for Political and Social Sciences* 604: 171–91.

Camillo, Charles A., and Matthew T. Pearcy. 2004. *Upon Their Shoulders: A History of The Mississippi River Commission from Its Inception Through the Advent of the Modern Mississippi River and Tributaries Project*. Vicksburg: Mississippi River Commission.

Campanella, Richard. 2007. "An Ethnic Geography of New Orleans." *Journal of American History* 94 (3): 704–15.

Colten, Craig E. 2005. *An Unnatural Metropolis: Wresting New Orleans from Nature*. Baton Rouge: Louisiana State University Press.

Colten, Craig E. 2009. *Perilous Place, Powerful Storms: Hurricane Protection in Coastal Louisiana*. Jackson: University Press of Mississippi.

Colten, Craig E. 2012. "Forgetting the Unforgettable: Losing Resilience in New Orleans." In *American Environments: Climate, Cultures, Catastrophes*, edited by Christof Mauch and Sylvia Mayer, 159–76. Heidelberg: Universitatsverlag.

"Court Overturns Liability." 1985. *New Orleans Times-Picayune*. April, 17, 1.

Day, John W. and Charles Hall, 2016. *America's Most Sustainable Cities and Regions: Surviving the 21st Century Megatrends*. New York: Springer.

"Everything You Need to Know About New Orleans Pumps, Plant That Helps Power Them." 2017. *New Orleans Advocate*, August 15. http://www.theadvocate.com/new_orleans/news/article_387a6d28-822e-11e7-8098-2b64c3d9c50c.html.

"Floods Not the Only Problem in Marrero Disaster Area." 1978. *New Orleans Times-Picayune*, May 12, 2.

"Here's Why New Orleans Flooding Was so Bad, and Why People Are Being Fired over It." 2017. *New Orleans Times-Picayune*, August 9. http://www.nola.com/politics/index.ssf/2017/08/floods_blame_sewerage_water_bo.html

Louisiana Office of Community Development—Disaster Recovery Unit. 2017. *Greater New Orleans Urban Water Plan*. http://livingwithwater.com/blog/urban_water_plan/about/

Melosi, Martin V. 2000. *The Sanitary City: Urban Infrastructure in America from Colonial Times to the Present.* Baltimore: John Hopkins University Press.

Meyer, William B., 2013. *The Environmental Advantages of Cities: Countering Commonsense Antiurbanism.* Cambridge, Mass.: MIT Press.

Mitchell, James. 1999. *Crucibles of Hazards: Mega-Cities and Disasters in Transition.* New York: United Nations University Press.

"Must Limit West Bank." 1978. *New Orleans Times-Picayune.* May, 10, 1 & 22.

NOAA. 2013. *National Coastal Population Report: Population Trends from 1970 to 2020.* Washington, D.C.: National Oceanographic and Atmospheric Administration.

NRC. 1999. *Our Common Journey: A Transition Toward Sustainability.* Washington, D.C.: National Academies Press.

O'Neill, Karen M. and Daniel J. Van Abs. 2016. *Taking Chances: The Coast After Hurricane Sandy.* New Brunswick, N.J.: Rutgers University Press.

Orleans Levee District. 1950. *Report on Flood Control and Shore Erosion Protection of City of New Orleans from Flood Waters of Lake Pontchartrain.* New Orleans: Orleans Levee District.

"Parishes Get Break in U.S. Flood Case." 1985. *New Orleans Times-Picayune.* April 9.

Rohland, Eleonora. 2019. *Hurricanes in New Orleans: Cultural Adaptation from the French Colonial Era to Hurricane Katrina.* New York: Berghahn Books.

Save Our Wetlands Inc. et al. v. Early J. Rush III et al. 1977. Injunction Order. United States District Court, Eastern District of Louisiana, Civil Action, 75–3710, December 30.

Save Our Wetlands Inc. et al. v. Early J. Rush III et al. 1978. Injunction Order. United States District Court, Eastern District of Louisiana, Civil Action 75–3710, March 10.

Seed, R. B., et al. 2005. *Preliminary Report on the Performance of the New Orleans Levee Systems in Hurricane Katrina on August 29, 2005.* Report No. UCB/CITRIS—05/01. Berkeley: University of California and ASCE.

Shallat, Todd. 2000. "In the Wake of Hurricane Betsy." In *Transforming New Orleans and Its Enviorns: Centuries of Change,* edited by Craig E. Colten, 121–138. Pittsburgh, PA: University of Pittsburgh Press.

Short, Lisa Benton. 2017. "Green Infrastructure, Green Space, and Sustainable Urbanism: Geography's Important Role." *Urban Geography.* DOI: 10.1080/02723638.2017.1360105.

Snowden, J. O., W. C. Ward, and J.R.J. Studlick. 1980. *Geology of Greater New Olreans: Its Relationship to Land Subsidence and Flooding.* New Orleans: New Orleans Geological Society.

Talen, Emily. 2011. "Sprawl Retrofit: Sustainable Urban Form in Unsustainable Places." *Environment and Planning B: Planning and Design* 38 (6): 952–78).

U.S. v. Jefferson Parish et al. 1987. "Settlement Agreement." New Orleans: U.S. District Court of the Eastern District of Louisiana, case nos. 81-1810 and 83-2077.

U.S. Army Corps of Engineers. 1962. *Interim Survey Report, Lake Pontchartrain Louisiana and Vicinity.* New Orleans, LA: U.S. Army Corps of Engineers, New Orleans District.

U.S. Army Corps of Engineers. 1966. *Hurricane Betsy, 8–11 September, 1965: After-Action Report.* New Orleans, LA: U.S. Army Corps of Engineers, New Orleans District.

U.S. Army Corp of Engineers. 1967. *Design Memorandum No. 2, General Design, Citrus.* New Orleans, LA: U.S. Army Corps of Engineers, New Orleans District.

U.S. Army Corps of Engineers. 1975. *Record of Public Meeting: Lake Pontchartrain, Loui-*

siana, and Vicinity, Hurricane Protection Project, February 22, 1975. New Orleans, LA: U.S. Army Corps of Engineers, New Orleans District.

U.S. Congress. 1950. *Lake Pontchartrain, Louisiana.* Letter from the Secretary of the Army, Senate Document 139, 81st Cong., 2nd sess.

Van Heerden, Ivor, et al. 2006. *The Failure of the New Orleans Levee System During Hurricane Katrina.* Baton Rouge: Louisiana Department of Transportation and Development.

Waggoner and Ball Architects. 2017. *Living with Water: A New Vision for Delta Cities.* http://livingwithwater.com/.

16

Suburban to Urban
Hydro-Economic Connectivity

Virtual Water Flow within
the Phoenix Metropolitan Area, Arizona

RICHARD R. RUSHFORTH AND BENJAMIN L. RUDDELL

Humans have become an urban species. Due to this, cities, the dominant habitat of humans, have become crucibles of natural resource consumption and a major driver environmental change—including through their supply chains. Therefore, we must examine and study how cities drive global environmental change (Grimm et al. 2008; Glaeser, Kolko, and Saiz 2001). City-induced environmental change is often measured through one of many types of environmental footprints (e.g., water, carbon, and ecological) that measure the pressure that humans put on natural resources and the environment via their supply chains. For example, a city's water footprint measures water consumed to produce food in distant agricultural areas, among many other types of direct and indirect water consumption. Here, we define direct water consumption as the consumption of water within a city for residents or economic production and indirect (virtual) water consumption as the volume of water consumed to produce a good or service that is consumed within the city. In the instance of food shipments from distant farms, this creates a transfer of virtual water from an agricultural area to the city (Hoekstra et al. 2012).

Metropolitan areas and megacities complicate this idealized conceptualization of the city because they contain diverse land uses and urban structures (e.g., natural lands, rural/agricultural, urban, suburban) that develop in response to social, environmental, and market forces (Lo and Yang 2002). These varied structures within metropolitan areas create a texture of economic niches that distinguish each urban subregion within the global economic network (Mills 1967). Metropolitan areas feature a rich pattern of dependencies between their constituent cities, in addition to depending on global supply chain connections.

Cities colocated within metropolitan areas simultaneously interact on multiple geographic scales, from the local to the global. At the local scale, metropolitan areas cities consume, directly and indirectly, a variety of natural and human resources. This consumption occurs via multiple shared infrastructure systems such as electric power, telecommunications, transportation, and water supply (Rinaldi, Peerenboom, and Kelly 2001; Pederson et al. 2006). The natural resources tend to be shared by multiple cities within a metropolitan area or region. A city's niche and structural role (e.g., core, suburban, exurban) is important due to competitive advantages and geographical relationships.

Water is an exemplar of a natural resource that is shared among a metropolitan area's constituent cities. Multiple cities tend to share natural systems–regional aquifers and watersheds–and engineered systems–transboundary aqueducts, water conveyance systems, and wastewater treatment facilities. Historically, these factors have necessitated regional water policies and plans. Regional policy is particularly important to govern rival but nonexcludable water resources (Roberts 1970; Davis 2007; Giordano and Wolf 2003; Rushforth, Adams, and Ruddell 2013; Ruddell et al. 2014). While direct water sharing agreements and policies reflect formal, long-term legal and political agreements, intrametropolitan virtual water flows reflect short-term economic conditions and competitive advantages. Both the long-term legal agreements allocating physical water resources, and the short-term trade agreements that imply virtual water transfers, influence intrametropolitan area hydro-economic decision-making on water infrastructure, investment, and operating costs.

When physical water supplies are scarce and shared, and constrain economic and demographic growth, cities can share water "virtually" by purchasing water-intensive goods and services from rural areas and other cities. This virtual water supply chain creates efficiencies, dependencies, tensions, and potentially conflicts. A virtual water supply chain that sources a city's food from a distant region does not create a conflict between neighboring cities. However, where cities are clustered in metropolitan areas virtual water sourcing can be local, and in competition with access to a shared physical water resource.

Therefore, each city in the metropolitan area must balance obtaining new water for economic growth, which excludes neighboring cities from using shared water resources, and outsourcing high-volume, low-value water uses to neighboring cities, which increases local water demand and pressure on the shared water resource. Local virtual water is a strategically important consideration for hydro-economic sustainability and resilience. While a city can outsource high-volume, low-value water consumption outside of its boundary, outsourcing to local virtual water sources does not reduce demand on shared water resources. We examine virtual water flows occurring within the Phoenix metropolitan area, a large metropolitan area with diverse land uses ranging from highly urbanized to rural agricultural, to understand indirect water sharing and virtual water dynamics in a water-constrained geography.

Case Study

In this chapter, we discuss virtual water flows that occur within the Phoenix metropolitan area (Rushforth and Ruddell 2015), a rapidly growing metropolitan area with shared, scarce water resources. The Phoenix metropolitan area is located in the Sonoran Desert in Arizona, USA, and has a population of 4.19 million people (U.S. Census Bureau 2011). Due to data availability, the study area was constrained to the 25 cities located in the conurbation surrounding the urban core of Phoenix, Mesa, Scottsdale, and Tempe (Zients 2013). These 25 cites have a combined population of 3.69 million people (Figure 16.1). For the cities in the Phoenix metropolitan area, the confluence of demographic and economic growth within an arid ecosystem has created competition between cities and economic sectors (industrial/commercial, residential, utilities, etc.) to secure water resources for future growth.

The major physical water sources for the Phoenix metropolitan area are the Colorado River, which Arizona accesses through the Central Arizona Project (CAP) and is shared with six other states and Mexico; the Salt and Verde Rivers via the Salt River Project (SRP); and a substantial, but nonrenewable, groundwater regional aquifer. The core Phoenix metropolitan area cities have greater access to surface water (the CAP and SRP systems), while smaller cities on the outskirts of the Phoenix metropolitan area are more dependent on groundwater (Sampson et al. 2011).

Within the Phoenix metropolitan area, virtual water flows can arise from both the commodity-producing and service sectors of the economy. Virtual water flow arising from the commodity-producing sectors of the economy were estimated using the National Water Economy Database (NWED) (Rushforth and Ruddell 2018). NWED was used to estimate

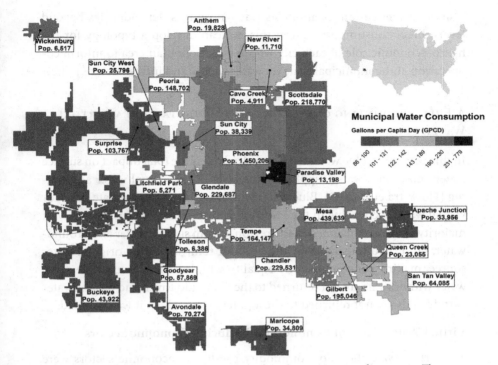

Figure 16.1. Potable water consumption in the Phoenix metropolitan area. The map above shows the population of the Phoenix metropolitan area cities included in the study boundaries along with residential GPCD for each city. Figure adapted from Rushforth and Ruddell (2015).

commodity-based virtual water flows at multiple scales from the intra-metropolitan area scale to the national scale. To date, no good method exists for estimating virtual water flow in the service sectors of the economy. However, since the service sectors are a large component of urban economies—71 percent of Phoenix metropolitan area residents are employed in the service sectors (U.S. Census Bureau undated)—we must develop novel methods to estimate these virtual water flows. For the Phoenix metropolitan area, we propose using labor flows between cities as an indirect indicator of virtual water flow in the service sectors of the economy. The service sectors of the economy do not consume water as an input like the commodity-producing sectors of the economy. Labor, via residential water consumption, represents the largest water consuming input to the service sectors of the economy, and can create intrametropolitan area water dependencies.

By evaluating both labor- and commodity-based virtual water flows at the metropolitan-area level, we can determine how the indirect consumption of a shared water resources amplifies water resources dependencies beyond levels direct consumption. From this, we can develop a typology for the hydro-economic role of each city within a metropolitan area to inform co-operation at the municipal scale on water supply and infrastructure policy.

A Full Accounting of Phoenix Metropolitan Area Virtual Water Flows

In this case study, blue water withdrawal, which is the total impact on surface water and groundwater supplies, is the basis of the virtual water accounting (Hoekstra et al. 2012). Blue water withdrawal is an appropriate basis for virtual water accounting in the Phoenix metropolitan area because the vast majority of water consumed in the region is from surface water and ground-water, not evapotranspiration, and also not returned to the original water source, making consumptive use equal to withdrawals. For example, CAP water withdrawals are not returned to the Colorado River, and groundwater withdrawals are not returned to the aquifers from which it was pumped.

Virtual Water Flows of Commodity-Producing Economic Sectors

The virtual water flows of commodity-producing economic sectors were estimated using NWED, which utilizes metropolitan area commodity flow data and county-level water withdrawal data (Southworth et al. 2010; Kenny et al. 2009). The NWED methodology has been described in detail by the authors in separate papers (Rushforth and Ruddell 2015; Rushforth and Ruddell 2018).

Estimating Virtual Water Flows of the Service Sectors of the Economy

Commuting is constrained by time, distance, and employment opportunities. Therefore, actual commute distance, travel time, and journey-to-work statistics for the Phoenix metropolitan area were used in this case study (Maricopa County 2011). Commuting flows between each city in the Phoenix metropolitan area labor flows were estimated using a network-based commuting flow model (Thorsen and Gitlesen 1998) constrained by total population and daytime population change data, ensuring that estimated commuting flows followed observed data (U.S. Census Bureau 2012; U.S. Census Bureau 2013).

Using commuting flow estimates, intrametropolitan and intramunicipal virtual water flows were calculated using city-specific gallon per capita day

(GPCD) and the commuting population between each Phoenix metropolitan area city.[1] For a detailed description of this methodology, please refer to the full journal article on Phoenix-area virtual water flows (Rushforth and Ruddell 2015).

Calculating a Complete Water Footprint Balance for the Phoenix Metropolitan Area

Using the commodity-based and labor-based methods for calculating virtual water flows, a net water footprint was calculated for each Phoenix metropolitan area city using the Embedded Resources Accounting (ERA) framework (Rushforth, Adams, and Ruddell 2013; Ruddell et al. 2014). ERA takes into consideration direct water withdrawals (U), which have been separated into potable deliveries (U_{urban}) and agricultural production (U_{farm}), and virtual water inflows (V_{In}) and outflows (V_{Out}) arising from each commodity (c) and labor flows (L) to arrive at a boundary-specific net water footprint (E) for a city (m). Virtual water inflows (V_{In}) are defined as the volume of water used to produce goods and services outside the municipal boundary but are consumed inside the municipal boundary; a subset of these flows are circular flows where production and consumption takes place within the same boundary. Virtual water outflows (V_{Out}) are defined as the volume of water used to produce goods and services that are consumed outside the municipal boundary. The equation applied to a Phoenix metropolitan area city's hydro-economy is as follows:

$$E_m = U_{urban} + U_{farm} + V_{C,In} - V_{C,Out} + V_{L,In} - V_{L,Out}$$

Each component of E_m can be further disaggregated to add granularity and specificity to the virtual water flows for each city. For example, to account for the multiple geographic boundaries that are analyzed simultaneously, V_{In} and V_{Out} terms are created for each of the other cities in the Phoenix metropolitan area (n) and each county in the United States (k). Labor flows to and from geographic areas outside of the Phoenix metropolitan area are considered to be negligible, so $V_{L,in}$ and $V_{L,out}$ are not considered for geographic areas outside the Phoenix metropolitan area. The following equation shows the specific form of the general ERA equation for each Phoenix metropolitan area city (m):

$$E_m = U_{urban} + U_{farm} + \Sigma_n(\Sigma_c V^n_{C,In} - \Sigma_c V^n_{C,Out} + V^n_{L,In} - V^n_{L,Out}) + \Sigma_k(\Sigma_c V^k_{C,In} - \Sigma_c V^k_{C,Out})$$

RESULTS AND DISCUSSION

National-Level Virtual Water Flows

At the national scale, virtual water inflows to Phoenix metropolitan area totaled 4,125 Mm^3 and virtual water outflows totaled 2,584 Mm^3. Virtual water inflows (V_{In}) were dominated by agricultural goods—processed foods, milled grain, animal feed, cereal grains. Our findings comport with numerous studies that previously identified the role that food plays in the global virtual water trade network (Hoff et al. 2014; Dalin et al. 2012; Konar et al. 2011; Mekonnen and Hoekstra 2011; Vanham and Bidoglio 2014). Further, virtual water inflows and outflows were dominated by local/regional trading partners: the Phoenix metropolitan area (first), Arizona (second), and Southern California (third). Overall, the Phoenix metropolitan area is a net importer of virtual water from the United States, with large virtual water flows originating from the Southwest U.S., especially Arizona, in addition other major metropolitan areas across the U.S.

Agricultural commodities originating from the western United States were the largest component of virtual water flows. In the western U.S., agriculture is primarily irrigated, unlike the eastern half of the United States where rainfall is more abundant and provides a greater proportion, if not all, of crop water demand. Thus, the geographic bias of Phoenix metropolitan area trade increases its virtual water intensity.

Virtual water flows associated with the trade of commodities averages 1,133 m^3 per capita for the Phoenix metropolitan area. Per capita virtual water outflows in the Phoenix metropolitan area follow a rough rank-order relationship from cities with high fractions of agricultural land (Buckeye) to residential/retirement cities (Sun City and Sun City West); ranging from 11,841 m^3 per capita in Buckeye to 3.0 m^3 per capita in Sun City West.

The western U.S. trade bias is due to dependence on commodities produced within the Phoenix metropolitan area. Approximately 30 percent of virtual water inflows to the Phoenix metropolitan area originates locally. The large dependency within the Southwestern U.S. region and Colorado River Basin amplifies exposure to water scarcity and potential disruptions to local water resource systems (Ruddell et al. 2014; Rushforth, Adams, and Ruddell 2013). Therefore, Phoenix metropolitan area virtual water trade amplifies dependency on shared local hydrology and physical water supplies rather than distributing water dependencies across hydrologically diverse regions.

Intrametropolitan Area Virtual Water Flows

Core cities are net virtual water importers from both metropolitan area neighbors and from cities and regions outside the Phoenix metropolitan area. Despite this overall pattern at the metropolitan area, agricultural cities within the Phoenix metropolitan area are net exporters of virtual water to both neighboring cities and the rest of the United States. While these results corroborate the results of numerous water footprint and urban metabolism studies that have found cities to be consumers of resources from distant regions, such as agricultural areas (Vanham and Bidoglio 2014), these results also show that in certain regions cities are also dependent on local hinterlands that share local water resources.

Small edge cities tended to have higher relative intrametropolitan virtual water flows and large, core cities had relatively higher levels of circular flows. The total virtual water flow associated with labor flows were 359 Mm^3. Core Phoenix metropolitan area cities—Phoenix, Scottsdale, Tempe—had the largest net inflow of labor-based virtual water. Further, half of the labor-based virtual water flow within the Phoenix metropolitan area was associated with the city of Phoenix. Following from this inflow of labor-based virtual water flow into core cities, suburban bedroom cities had the largest net labor-based virtual water outflow. On average, 36 percent of labor-based virtual water flows were due to commuting from suburban cities to core cities. The remaining 64 percent of labor-based virtual water flows were circular flows within each city, but this intracity virtual flow was dominated by Phoenix. These results provide evidence that Phoenix metropolitan area commuting does not just follow a suburban-to-central city pattern; rather, the metropolitan area has a decentralized and poly-nucleated commuting pattern (Baum-Snow 2010).

Since Phoenix metropolitan area cities share common physical water resources, the net flows of virtual water within the metropolitan area are conceptually interchangeable with a proportionate physical reallocation of physical water resources. The high degree of intra-Phoenix metropolitan area virtual water flows further reveals the magnitude of physical water resource sharing. Within the Phoenix metropolitan area, virtual water flows are hydro-economic relationships between independently managed, but mutually dependent, water distribution systems.

The relative magnitude of the virtual reallocation of water is estimated by the comparison between the direct water withdrawals (U) and the intrametropolitan net water footprint of each city (E_{PMA}) (Figure 16.2). Core cities

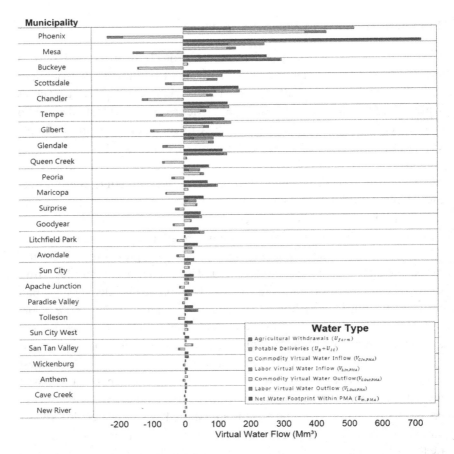

Figure 16.2. Phoenix metropolitan area city water footprint components. Cities have different roles in the metropolitan economy. Core cities tend to have virtual water inflows that are greater than outflows and potable system deliveries. Bedroom cities have greater outflows of virtual water associated with labor than corresponding inflows. The net water footprint (E) is a city's total impact on the Phoenix metropolitan area's shared physical water resources. Figure adapted from Rushforth and Ruddell (2015).

have a larger water footprint on the area's shared physical water resources when virtual water flows within the metropolitan area are considered; the opposite is true for edge and bedroom cities. Core cities depend disproportionately on the water supplies of neighbors rather than the water supplies of more distant trading partners.

A General Hydro-Economic Typology for Communities

A generalized hydro-economic typology can be created based on the relative virtual water flows between Phoenix metropolitan area cities using commodity-based and labor-based virtual water flows.[2] Within the Phoenix metropolitan area there are at least four qualitatively different hydro-economic types of cities: (1) core cities which are high-value economic centers and job centers that are dependent on their neighbors for net virtual water inflows in both labor and commodities; (2) suburban bedroom cities that are net virtual water exporters to core cities via labor flows but net virtual water importers of commodities because of their relatively large residential populations (Kenessey 1987); (3) edge cities that are net virtual water exporters, especially of agricultural commodities but also of other commodities and labor; and (4) transitional core cities, which have become job centers and are therefore net importers of virtual water in labor, but are still net exporters of commodities, possibly due to economic specialization in an area such as manufacturing, or due to significant remaining agricultural activity. A balanced city is neither a significant virtual water importer or exporter. This balance might be because the community has equal parts of each of the four types described above, or because the community is so small that it trades very little (Figure 16.3).

Conclusions

The high likelihood of drought in the Southwest (Cook, Ault, and Smerdon 2015) poses challenges to the Phoenix metropolitan area economy, the local water resources system multiple scales, and regional water resources management (Gober and Kirkwood 2010). From a hydro-economic perspective, the 25 cities of the Phoenix metropolitan area function as one interdependent city. While each city within the Phoenix metropolitan area may individually plan for drought and long-term water scarcity, the effectiveness of drought planning is manifest at the metropolitan area scale, not the city. The impacts of water rationing, curtailment of water supply, or the failure of water infrastructure goes beyond the city to the entire metropolitan area hydro-economy. Core cities tend to have strong water rights positions and are much more insulated from the effects of drought than the bedroom and edge cities on which they are hydro-economically dependent. In view of likely drought, it might benefit Phoenix metropolitan area cities to pursue infrastructure and policy that recognizes this fact.

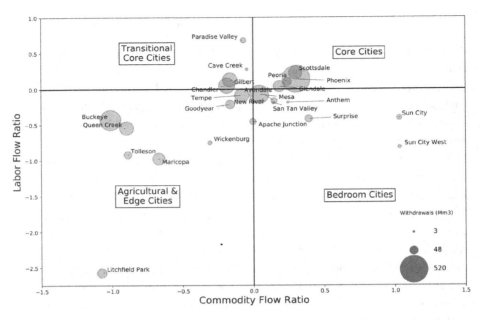

Figure 16.3. Phoenix metropolitan area city hydro-economic typology. The Phoenix metropolitan area's leading cities, Phoenix and Scottsdale, typify the core community, and heavily agricultural cities such as Queen Creek and Buckeye typify the edge community. Chandler and Gilbert are transitional core cities that are developing to resemble Scottsdale but are currently part agricultural. Tempe and Mesa are balanced hydro-economies. This typology is based only on intrametropolitan virtual water flows and describes the relative hydro-economic role of each city within the metropolitan area. Figure reproduced with data from Rushforth and Ruddell (2015).

One strategy for Phoenix metropolitan area cities to enhance hydro-economic sustainability and resilience is to pursue public/private policies that increase virtual water sourcing from less drought-prone and less water-stressed geographies—in particular, outsourcing high-volume, low-value water consumption to water abundant areas while retaining and growing high-value water consumption. This strategy adds an indirect supply chain component to complement the traditional approach to urban water supply policy that emphasizes water efficiency and multiple redundant physical water sources. Another strategy is for core cities to more actively cooperate with bedroom and edge cities on issues of water rights, water infrastructure investment, and water allocation policy to ensure that the entire metropolitan area is hydro-economically secure.

This second strategy is more promising in metropolitan areas like the Phoenix metropolitan area where much of the water supply chain of core cities is locally sourced. However, it carries the political and hydrological consequence of increasing the dependency of core cities on bedroom communities. If there is water scarcity, regional city managers and elected city council members will be faced with the tradeoff between maintaining hardened residential water demand and maintaining water for economic production, and potential backlash from voters.

Each city type within a metropolitan area has a distinct point of view with respect to cooperative water policy and may choose to acknowledge or discount the indirect component of the intrametropolitan water footprint. Core cities benefit the most from positive externalities and a lower apparent water footprint by neglecting the indirect dependency and are less likely to see that cooperation with other cities on water infrastructure investment is in their best interest. Edge cities have the strongest interest in adopting a complete water footprint balance because they are important providers of water-derived goods and services and have a net water footprint that is lower than is at first apparent. However, edge cities are the most vulnerable to disruptions in water supply due to their limited economic and political power and their relatively water-intensive economies. Because core cities depend on edge cities, there is a shared interest in using this information to guide cooperative water policy and investment.

Intrametropolitan area virtual water flows (suburban to urban) are fundamentally different from mesoscale and macroscale virtual water flows because they can directly substitute for physical water flows (Merrett 2003). Intrametropolitan area virtual water flows occur at hydrologically colocated scales where infrastructure and water rights divide the physical water resource into multiple separate stocks. These multiple water stocks can suffer from different levels of stress, scarcity, or disruption due to differences in investment and water rights, rather than hydrological differences. Therefore, at the metropolitan-area level disruption risks are the direct result of water policy, law, and investment, and can be addressed through virtual water cooperation between core cities, transitional core and bedroom (suburban) cities, and edge/agricultural (exurban) cities within metropolitan areas.

While the framework presented in this chapter has utilized water and virtual water as a case study, the framework is readily extensible to other shared natural resources at the metropolitan-area level in addition to regional greenhouse gas emissions management.

Acknowledgments

The case study presented in this chapter is a condensed and updated version of a study previously published in *Sustainability*. For detailed results and methodology, please refer to the original journal article: Rushforth, R. Richard, and L. Benjamin Ruddell. 2015. "The Hydro-Economic Interdependency of Cities: Virtual Water Connections of the Phoenix, Arizona Metropolitan Area." *Sustainability* 7(7): 8522–8547.

Additionally, the authors acknowledge funding from the U.S. National Science Foundation via the Central Arizona-Phoenix Long-Term Ecological Research (CAP-LTER, CAP3) grant BCS-1026865, and the collaborative support of the National Water-Economy Project (NWEP). The authors acknowledge helpful input from Mr. Doug Toy of the City of Chandler, Arizona. The views expressed are those of the authors, and not the funding agencies or acknowledged collaborators.

Notes

1. GPCD data for each Phoenix metropolitan area city were obtained from the Arizona Department of Water Resources (https://infoshare.azwater.gov/docushare/dsweb/HomePage). Data are reported by cities in GPCD and Loss and Unaccounted (L&U) for Water Percentage reports. City and report numbers in parentheses are as follows: Town of Apache Junction (56–002025.0000); City of Avondale (56–002003.0000); Town of Buckeye (56–002006.0000 and 56–002012.0000); Town of Cave Creek (56–002008.0000); City of Chandler (56–002009.0000); Town of Gilbert (56–002017.0000; City of Glendale (56–002018.0000); City of Goodyear (56–002019.0000; Litchfield Park (56–002021.0000); Town of Maricopa (56–001355.0000); City of Mesa (56–002023.0000); Town of New River (56–002254.0000); Town of Paradise Valley (56–002027.0000); City of Peoria (56–002029.0000); City of Phoenix (56–002030.0000); Town of Queen Creek (56–002032.0000); San Tan Valley (56–002020.0000); City of Scottsdale (56–002037.0000); Town of Sun City (56–002038.0000); Sun City West (56–002039.0000); City of Surprise (56–002344.0000); City of Tempe (56–002043.0000); and City of Tolleson (56–002044.0000). Data for the Town of Wickenburg were obtained from that town's annual comprehensive financial reports. Commuting data—population, mode, and distance—were obtained from the Maricopa County Air Quality Department's Trip Reduction Program Annual Report, which surveys journey to work behavior in Maricopa County, Arizona.

2. We use a Labor Flow Ratio (LFR), defined as [EQUATION 1] and a Commodity Flow Ratio (CFR), defined as [EQUATION 2] metrics for each Phoenix metropolitan area city to create this hydro-economic typology.

REFERENCES CITED

Baum-Snow, Nathaniel. 2010. "Changes in Transportation Infrastructure and Commuting Patterns in US Metropolitan Areas, 1960–2000." *American Economic Review* 100(2): 378.

Cook, Benjamin I., Toby R. Ault, and Jason E. Smerdon. 2015. "Unprecedented 21st Century Drought Risk in the American Southwest and Central Plains." *Science Advances* 1(1): e1400082.

Dalin, Carole, Megan Konar, Naota Hanasaki, Andrea Rinaldo, and Ignacio Rodriguez-Iturbe. 2012. "Evolution of the Global Virtual Water Trade Network." *Proceedings of the National Academy of Sciences* 109(16): 5989–5994.

Davis, Matthew D. 2007. "Integrated Water Resource Management and Water Sharing." *Journal of Water Resources Planning and Management* 133(5): 427–445.

Giordano, Meredith A., and Aaron T. Wolf. 2003. "Sharing Waters: Post-Rio International Water Management." Natural Resources Forum 27(2): 163–171.

Glaeser, Edward L., Jed Kolko, and Albert Saiz. 2001. "Consumer City." *Journal of Economic Geography* 1(1): 27–50.

Gober, Patricia, and Craig W. Kirkwood. 2010. "Vulnerability Assessment of Climate-Induced Water Shortage in Phoenix." *Proceedings of the National Academy of Sciences* 107(50): 21295–21299.

Grimm, Nancy B., Stanley H. Faeth, Nancy E. Golubiewski, Charles L. Redman, Jianguo Wu, Xuemei Bai, and John M. Briggs. 2008. "Global Change and the Ecology of Cities." *Science* 319(5864): 756–760.

Hoekstra, Arjen Y., Ashok K. Chapagain, Maite M. Aldaya, and Mesfin M. Mekonnen. 2012. *The Water Footprint Assessment Manual: Setting the Global Standard.* London, UK: Earthscan.

Hoff, Holger, Petra Döll, Marianela Fader, Dieter Gerten, Stefan Hauser, and Stefan Siebert. 2014. "Water Footprints of Cities—Indicators for Sustainable Consumption and Production." *Hydrology and Earth System Sciences* 18(1): 213–226.

Kenessey, Zoltan. 1987. "The Primary, Secondary, Tertiary and Quaternary Sectors of the Economy." *Review of Income and Wealth* 33(4): 359–385.

Kenny, Joan F., Nancy L. Barber, Susan S. Hutson, Kristin S. Linsey, John K. Lovelace, and Molly A. Maupin. 2009. *Estimated Use of Water in the United States in 2005.* Reston, VA: U.S. Geological Survey.

Konar, M., C. Dalin, S. Suweis, N. Hanasaki, A. Rinaldo, and I. Rodriguez-Iturbe. 2011. "Water for Food: The Global Virtual Water Trade Network." *Water Resources Research* 47(5). https://doi.org/10.1029/2010WR010307.

Lo, C. P., and Xiaojun Yang. 2002. "Drivers of land-use/land-cover Changes and Dynamic Modeling for the Atlanta, Georgia Metropolitan Area." *Photogrammetric Engineering and Remote Sensing* 68(10): 1073–1082.

Maricopa County Air Quality Department. 2011. *Trip Reduction Program Annual Report 2010.* Phoenix, AZ: Maricopa County Air Quality Department.

Mekonnen, Mesfin M., and Arjen Y. Hoekstra. 2011. National Water Footprint Accounts: The Green, Blue and Grey Water Footprint of Production and Consumption. Delft, the Netherlands: Unesco-IHE Institute for Water Education.

Merrett, Stephen. 2003. "Virtual Water and the Kyoto Consensus." *Water International* 28(4): 540–542.

Mills, Edwin S. 1967. "An Aggregative Model of Resource Allocation in a Metropolitan Area." *American Economic Review* 57(2): 197–210.

Pederson, Peter, D. Dudenhoeffer, Steven Hartley, and May Permann. 2006. *Critical Infrastructure Interdependency Modeling: A Survey of US and International Research.* Technical Support Working Group. Idaho Falls: Idaho National Laboratory.

Rinaldi, Steven M., James P. Peerenboom, and Terrence K. Kelly. 2001. "Identifying, Understanding, and Analyzing Critical Infrastructure Interdependencies." *Control Systems, IEEE* 21(6): 11–25.

Roberts, Marc J. 1970. "River Basin Authorities: A National Solution to Water Pollution." *Harvard Law Review* 83(7): 1527–1556.

Ruddell, Benjamin L., Elizabeth A. Adams, Richard Rushforth, and Vincent C. Tidwell. 2014. "Embedded Resource Accounting for Coupled Natural-Human Systems: An Application to Water Resource Impacts of the Western US Electrical Energy Trade." *Water Resources Research* 50(10): 7957–7972.

Rushforth, R. R., and B. L. Ruddell. 2018. "A Spatially Detailed Blue Water Footprint of the United States." *Hydrology and Earth System Sciences* (22)5: 3007–3032. https://doi.org/10.5194/hess-22-3007-2018.

Rushforth, R. Richard, and L. Benjamin Ruddell. 2015. "The Hydro-Economic Interdependency of Cities: Virtual Water Connections of the Phoenix, Arizona Metropolitan Area." *Sustainability* 7(7): 8522–8547.

Rushforth, Richard R., Elizabeth A. Adams, and Benjamin L. Ruddell. 2013. "Generalizing Ecological, Water and Carbon Footprint Methods and Their Worldview Assumptions Using Embedded Resource Accounting." *Water Resources and Industry* 1–2 (March–June): 77–90.

Sampson, D. A., V. Escobar, M. K. Tschudi, T. Lant, and P. Gober. 2011. "A Provider-Based Water Planning and Management Model—WaterSim 4.0—for the Phoenix Metropolitan Area." *Journal of Environmental Management* 92(10): 2596–2610.

Southworth, Frank, Diane Davidson, Ho-Ling Hwang, Bruce Peterson, and Shih-Miao Chin. 2010. *The Freight Analysis Framework, Version 3: Overview of the FAF3 National Freight Flow Tables.* Washington, D.C.: U.S. Department of Transportation.

Thorsen, Inge, and Jens Petter Gitlesen. 1998. "Empirical Evaluation of Alternative Model Specifications to Predict Commuting Flows." *Journal of Regional Science* 38(2): 273–292.

U.S. Census Bureau. 2011. "Population Change for Metropolitan and Micropolitan Statistical Areas in the United States and Puerto Rico: 2000 to 2010." Last accessed December 2, 2014. https://www2.census.gov/programs-surveys/decennial/tables/cph/cph-t/cph-t-2/cph-t-2.xls.

U.S. Census Bureau. 2012. "Table 3. Commuter-Adjusted Daytime Population: Places." Last accessed December 2, 2014. https://www2.census.gov/programs-surveys/commuting/tables/time-series/acs2006-2010/table3.xls.

U.S. Census Bureau. 2013. *Summary of Population and Housing Characteristics.* 2010 Census of Population and Housing. Last accessed December 2, 2014. https://www2.census.gov/library/publications/2012/dec/cph-1-1.pdf.

U.S. Census Bureau. Undated. "LED Extraction Tool—Quarterly Workforce Indicators (QWI)." Last accessed February 17, 2020. https://qwiexplorer.ces.census.gov/static/explore.html#x=0&g=0.

Vanham, Davy, and G. Bidoglio. 2014. "The Water Footprint of Milan." *Water Science & Technology* 69(4): 789–795.

Zients, Jeffrey. 2013. *Revised Delineations of Metropolitan Statistical Areas, Micropolitan Statistical Areas, and Combined Statistical Areas, and Guidance on Uses of the Delineations of These Areas.* OMB Bulletin No. 13-01. Washington, D.C.: Office of Management and Budget.

17

Conclusions

Suburban Sustainability Themes and Ongoing Challenges

ROBERT BRINKMANN AND SANDRA J. GARREN

Sustainability initiatives in the United States at this particular moment in time are extremely challenging due to the lack of leadership in the executive branch of the federal government and an absence of cooperation and agreement in the legislative branch (Brewer 2014). As a result, there are some states and local governments that have emerged as leaders in this area (Watts 2017). While some large cities in the United States and throughout the world have stepped forward as leaders on sustainability (Mohareb, Heller, and Guthrie 2018), few have recognized the contributions and significance of the suburbs as worthy of attention. Places like New York City and Burlington, Vermont, in part through the efforts of former mayors Michael Bloomberg and Bernie Sanders and other strong leaders, have done amazing things to advance sustainability in very big ways. It is important to note that we argue that sustainability goals in cities are far easier to accomplish than in the suburbs. Cities benefit from a strong top-down local government, rich treasuries (in some cases), and an environmentally conscious population. The suburbs, in contrast, often have diffused and weak local governments, small budgets, and an economically frugal and less environmentally conscious population (Hamel and Keil 2015). Doing sustainability in the suburbs is hard.

Even though it is difficult, sustainability initiatives in these low population density landscapes are important. If one takes a look at most metro regions in the United States, it is clear that the spatial extent of suburbs far exceeds the footprint of most cities (see Figure 1.2). Chicago, for example, consist of 234 square miles of land. However, its metro region, which is largely suburban, consists of 4,198 square miles (Figure 17.1). Spatially, the

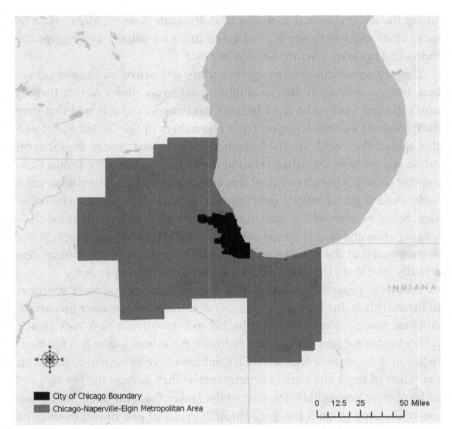

Figure 17.1. Urban and suburban land map in the Chicago-Naperville-Elgin metropolitan area.

decisions impacting the land and people can have tremendous impacts in the suburbs.

Of course, the focus on sustainability in cities aims at reducing the environmental footprint of individuals by addressing big infrastructure issues. Cities take on large sustainability problems like energy or water consumption, transportation, environmental justice, and green economic development in highly organized fashion. Through particular programs city governments are able to reduce the footprint of citizens. New York did not reduce its environmental footprint by relying on individual citizens to make decisions, they did it through the efforts of government offices that were working on sustainability goals. This is very different from the suburbs where the environmental footprint of individuals is much higher than in most cities and where any type of green initiative usually will engage the individual in closer ways than found in cities.

The large environmental footprint and the expansive land use of suburban dwellers relative to their neighboring urbanites shows us that there is much needed work to be done in these landscapes, and it is odd that they have received so little attention from researchers in the United States and throughout the world. This lack of attention is, in part, due to the focus on urban studies after the urban crises in the United States in the 1960s. During that decade it became evident that cities had many problems associated with white flight, economic development, infrastructure, and social cohesion. Scholars interested in human settlement rightly focused on these challenging areas and largely ignored the changing suburban landscapes. While some noted that the suburbs were a growing environmental problem, few actually gave them the attention they deserved as they expanded.

Today, the suburbs have matured and scholars are paying more attention to them. This is due partially to the fact that the problems once focused in cities are now present in the suburbs. For example, in the New York area, it is far cheaper for new immigrants to live in the suburbs than it is for them to live in most areas of New York City. Suburbs have never had to deal with the influx of large numbers of immigrants as they have in the last two decades (Lowenhaupt 2016). The days of the Lower East Side as an immigrant mecca are long gone; it has gone through cycles of gentrification since its time as a focal point of immigrant settlement (Timberlake and Johns-Wolfe 2016). Today's immigrants are not choosing to go to areas near downtown New York, but instead move to suburban communities in places like Long Island and Connecticut. Immigration is just one way the suburbs have changed. However, the suburbs are also dealing with emerging issues with public health such as the opioid epidemic (Hansen 2017); racial tensions such as was seen in Ferguson, Missouri, in suburban St. Louis (Derickson 2016); and serious infrastructure problems such as the groundwater pollution from septic systems as experienced in suburban Tampa (van Beynen et al. 2012). These are but a few of the challenges that suburbs are facing as they age.

KEY THEMES

As the suburbs confront their challenges, some innovative regions are using sustainability as a framework for positive change. This book highlights how individual places and regions are moving the sustainability agenda forward in the suburbs. Through the work of local governments, nonprofit groups, and dedicated individuals, suburbs are addressing many of the key sustainability initiatives that the United States faces. This is the first book to pull together some of the innovative sustainability ideas and initiatives present within the suburban landscape of the United States. It is clear from this work that there are three main themes that emerge.

Cooperation across local governments and regions leads to success. Suburban governments are highly fragmented with a diversity of political outlooks. Some local governments are extremely frugal and nearly libertarian in outlook while others are extremely liberal with public spending. As a result, regional suburban initiatives are difficult to undertake without some regional body providing a framework for cooperation. These regional bodies may take the form of a loose broad regional governance such as a regional planning body, or it may take the form of a thematic regional governance in the form of something like a water, energy, or transportation authority. These organizations have the ability to develop programs that help suburbs with disparate political outlooks reach sustainability goals. Cooperation of local governments is essential in order to reach targets. Working with local and regional stakeholders in such settings can be challenging, but it is clear from the examples presented in this book that positive outcomes are possible. For example, Dierwechter in chapter 3 highlighted a successful example of regional and local planning targeted in a suburban community in greater Seattle, Washington, and Mattiuzzi in chapter 11 noted the significance of sustainability governance to promote sustainability goals in the Los Angeles, California, region.

Transportation and other infrastructure needs to be rethought. One of the major problems associated with suburbs is that they rely on a diffuse and inefficient infrastructure. Transportation is the most obvious problem. However, many other infrastructure issues exist. Intensive road, water, and energy infrastructure in low population density areas makes them very expensive to build and maintain. However, some suburban areas are trying to change things. For example, in chapter 5, Garren and others discuss the major roadway changes taking place in one community in Long Island, New York, that is part of a national effort to make the suburbs not only more sustainable, but also safer. Similar efforts are also under way in Tampa, but under the aegis of brownfields redevelopment as outlined by Wells and others in chapter 7.

Measuring and assessment are integral to success. Doing local or regional sustainability is all about benchmarking against standard operating procedures and finding ways to quantitatively improve conditions to reduce the environmental footprint of a population. While this is regularly done in cities, states, and even countries, scant information on measuring sustainability in suburbs is in the literature. This body of work demonstrates that sustainability measurements and assessments are being done in the suburbs and that these initiatives lead to improvements in suburban sustainability. It is important to stress that many of the suburban measurements and assessments are being conducted as part of regional assessments of sustainability, but they are being done. Thus, the suburbs should not be written off as somehow too difficult to benchmark or assess. They are an important part of the settlement fabric of the United States and we should find ways to assess their role in the future sustainability of the nation by finding ways to measure and benchmark basic sustainability metrics. Some examples from this book show that benchmarking is important. For example, Garren in chapter 2 demonstrates how an index can be used to assess suburban sustainability within a regional context in the New York metropolitan region, and Rushforth and Ruddell in chapter 16 note the significance of benchmarking water flows within the Phoenix metro.

Individual governments or actors matter. Sustainability is not on the radar of most small suburban governments. They are more concerned with local issues such as roads, schools, zoning, and local taxes. To many small governments, sustainability is something that others are dealing with at the state, national, or international level. However, individual actors, which can include local activists, elected officials, or business leaders, push local governments to act on important sustainability initiatives that address one or more aspects of the environmental, social, and economic triad of sustainability. While nations or states may have a national sustainability agenda, the suburbs do not. Thus, it is often the leaders in suburbs that provide the impetus for creating positive change. While the initiatives developed by these actors is important, this issue highlights that there is a lack of national coordination in the suburbs to create state or national suburban sustainability goals or initiatives. Certainly organizations are present to help suburbs in their sustainability initiatives. For example, the U.S. Green Building Council provides LEED ratings for cities and communities. On Long Island, Vision Long Island provides support and guidance for suburban communities interested in implementing Smart Growth initiatives. There are many more local initiatives around the nation that assist suburbs in particular areas. But local governments may or may opt out of these initiatives, and local government leadership changes. Thus, it is communities where there are individual actors or strong groups working on sustainability where the most progress is made. Unfortunately, the lack of a coordinated effort means that the work in suburbs is uncoordinated, not regionally integrated, and diffuse. Much more work needs to be done to try to coordinate suburban sustainability goals. Some areas have worked hard to achieve interesting results. For example, the maker movement in Georgia as highlighted by Opp in chapter 8 provides a clear example of the significance of local institutions in promoting sustainable economic development, and the work of Wells and colleagues in chapter 7 highlights the importance of community engagement in helping to engage diverse populations in suburban community design.

CONTINUING CHALLENGES AND OPPORTUNITIES

While great things are happening in the suburbs in the area of sustainability, it is important to stress that there are continuing challenges. Several are highlighted below. They broadly relate to infrastructure and resources. However, it is worth stressing that there is also a paucity of research on this very important landscape. Thus, one of the biggest challenges faced in suburban sustainability is a lack of information or assessment. The focus for most sustainability experts working in regions is within cities. While this is changing, the academic community should work harder to integrate suburbs into the sustainability research agenda. Below is a list of key challenges facing suburbs in the area of sustainability and the opportunities present to try to address these challenges.

Settlement patterns. There is no doubt that the basic settlement patterns of suburbs are problematic for sustainability. Low population density, lack of walkability, and street patterns are but a few of the issues built into the fabric of suburban communities. While it is difficult to change these patterns, some suburban communities are working hard to address these issues through densification programs, rethinking street patterns, and improving walkability and bike access. Nevertheless, the basic design of the suburbs remains a challenge. While efforts can be made to change the design of the suburbs, it is important to try to improve the sustainability of the suburbs as they exist. Thus, even with the problematic settlement patterns, one can improve various sustainability indicators such as energy use, water consumption, environmental justice, and so forth within the existing system.

Transportation, water, and energy infrastructure. Many suburban communities have problematic infrastructure challenges with transportation, water and energy. Many were built without access to public drinking water or sewage systems, and many were built far from power plants making them vulnerable to power grid issues. Some areas have seen serious water pollution and water shortage problems. And some places have had serious energy issues emerge due to forest fires in the suburban West. Improving transportation, energy, and water infrastructure in suburbs is expensive and thus a difficult proposition since there is a limited tax base. Yet there are opportunities, particularly in the area of energy. Local green energy production is becoming much more common in the suburbs. Whether it is an individual house or a small wind or solar plant, many suburban communities are becoming much more resilient from problems associated with large regional energy production. New, innovative water planning and governance groups, like the Edwards Aquifer Authority in central Texas, confront regional water issues that impact not only cities, but the surrounding suburbs and rural communities. In addition, new approaches to suburban redevelopment are improving roadways and making them safer.

Economic patterns. For decades the suburbs have been seen as the bedroom communities of large cities. The suburbs provide the workers for the urban economic engine. This is problematic in a sustainability lens due to the transportation issues associated with this pattern. Large roads need to be built and there are massive energy costs associated with bringing people to and from their urban jobs from the suburbs. Until recently, the United States hasn't really looked at the suburbs as areas of major job growth. People moved out to the suburbs to get away from the industrial and commercial activities present in cities. This is clearly changing in the United States. Suburban areas are much more comfortable with attracting a variety of economic activities to attract jobs for their citizens. Small industrial parks and commercial and financial centers are all welcome in most suburbs today. In addition, the suburbs are becoming much more friendly to small agricultural developments. Small commercial and nonprofit farms are increasing across the suburban landscape. Many communities have changed their zoning to allow beekeeping and backyard chickens. While the suburbs are still supplying many workers to the cities, there are more economic options in the suburbs than there have been. More work on this needs to advance to try to provide greater opportunity for suburban residents.

Environmental justice. The basic tenet of environmental justice is that everyone in a community shares in the benefits and risks associated with environmental problems in a community. Suburbs have particular challenges in this area. We have many people who have moved to the suburbs to get away from particular environmental problems imposed upon them. At the same time, there are many suburbs that were developed and marketed on lands that have old remnant environmental problems such as the presence of hazardous waste or new problems such as a new power plant. Due to the relatively small size of suburban local governments, these problems can manifest in existential ways for communities. At the same time, the suburbs in the United States are becoming much more diverse. In part because of this diversity, the suburbs are becoming much more aware of environmental justice. Indeed, many small communities have become the focus of environmental justice efforts in recent years. Recent work focused on diverse suburban communities in Louisiana (Frankland and Tucker 2015) demonstrates that while there are many remaining problems, suburban activism can make a difference in the lives of residents.

Climate change and natural hazards. It is clear that the suburbs have been and will continue to be impacted by climate change and natural hazards. For example, when Superstorm Sandy hit the New York City region in 2012, the suburban areas of the region had a difficult time recovering (Ploran, Trasciatti, and Farmer 2018). New York City was certainly badly damaged, but the suburbs didn't have the resources to address a number of problems—from loss of homes to loss of infrastructure. However, there is more attention being given as of late to suburban resilience (Sanderson et al. 2016). Suburban regions are engaging on issues of natural hazards and climate change in new ways. They are working with regional partners on innovative projects to make their communities more resilient to the impacts of climate change and natural disasters.

Looking Beyond the United States

This book focused exclusively on the suburbs of the United States. The reason for this is mainly because United States suburbs are rather different from suburbs in other parts of the world. American suburbs tend to be less dense and less connected to regional governance and mass transportation networks. But it is worth considering what is occurring in other parts of the world where suburbs impact national sustainability initiatives.

Suburban development in many other parts of the world can be divided into two broad categories, planned and unplanned. Planned suburbs, such as New Cairo south of Cairo and Shunyi near Beijing, have many elements of the new urbanism movement: pockets of density, open parks, and commercial centers. Infrastructure is built into the suburbs, and they have distinct local character. In contrast, unplanned suburbs such as seen in areas on the fringes of Rio de Janeiro or Mumbai develop legally or illegally in relatively haphazard fashion. These suburbs typically do not have significant infrastructure and have a very high population density. They are usually not connected to local governance and are often in conflict with governments due to the challenging issues associated with them. They often become edge slums that provide particular challenges for infrastructure improvement. Compared to American suburbs, much has been written about these two categories of suburban development throughout the world from a sustainability lens. Many of the planned communities have been built with sustainability in mind. For example, in India, where there are significant challenges due to unplanned suburbs, there are efforts under way to focus suburban developing using smart city sustainability metrics (Randhawa and Kumar 2017). In Europe, where land for development is tightly controlled, most suburbs are built with very strong planning guidelines. Sustainability has infused much of the suburban development within the last two decades, particularly as the European Union provided sustainability guidelines and goals. In contrast, in Latin America and Africa, many local and national governments are trying to confront the economic, social, and environmental challenges that have emerged due to unplanned suburban development (Cobbinah and Aboagye 2016; Jaitman 2015).

While these two categories of suburbs found throughout the world, planned and unplanned, are of interest, they are distinctly different from the car-focused suburbs of the United States. Because American suburbs developed so distinctly differently from suburbs in many other parts of the world, they are worthy of their own discussion. A recent phenomenon, the development of American-style suburbs in the developing world, has received attention as of late (Klaufus et al. 2017). There is concern that the car-focused developments popping up around the world will have negative impacts on national and regional sustainability initiatives. Plus, the advent of the gated community seems to be taking hold in some parts of the world where there are significant issues with income equality.

CREATING A SUSTAINABLE SUBURB

Of course, as Hanlon (2017) aptly notes, the suburbs in the United States are rather diverse, and they are changing quickly. The typical American has a range of patterns that influence overall suburban sustainability. Yet there is much more focus on how we work to create sustainable suburbs. It is important to note, however, there are many communities that are moving forward with a variety of sustainability initiatives that help regions achieve their sustainability goals. Something is going on with sustainability in every corner of the United States. For example, in Florida, in the Villages, an upscale retirement village, residents can travel to commercial zones in the complex on golf carts (Bartling 2006) and in suburban Long Island, old suburbs like Hicksville are undergoing significant redevelopment initiatives to refocus on densification and downtown development. While many of these initiatives are not comprehensive across all areas of sustainability as one might see in coordinated city initiatives, they are evidence that sustainability is infusing suburbia.

In the coming years, we predict we will see greater emphasis on smart growth in planned suburbs. In addition, there will be considerable attention on transforming suburbs into more sustainable communities through redevelopment and densification initiatives. In addition, many unique themes are developing in the area of suburban sustainability that range from wildlife management to suburban agriculture, and from innovative energy projects to water management. While it is unlikely that American suburbs will ever have a per capita environmental footprint that is less than their neighboring cities, there are tremendous opportunities to make real quantitative advances that will help the United States to become a more sustainable nation.

References Cited

Bartling, H. 2006. "Tourism as Everyday Life: An Inquiry Into The Villages." *Florida, Tourism Geographies* 8: 380–402. DOI: 10.1080/14616680600922070.

Brewer, T. L. 2014. *The United States in a Warming World: The Political Economy of Government, Business, and Public Responses to Climate Change.* Cambridge: Cambridge University Press.

Cobbinah, P. B., and Aboagye, H. N. 2016. "A Ghanaian Twist to Urban Sprawl." *Land Use Policy* 61: 231–241.

Derickson, K. D. 2016. "Urban Geography II: Urban Geography in the Age of Ferguson." *Progress in Human Geography* 41: 230–244.

Frankland, P., and S. Tucker. 2015. *Women Pioneers of the Louisiana Environmental Movement.* Jackson: University Press of Mississippi.

Hamel, P., and R. Keil, eds. 2015. *Suburban Governance: A Global View.* Toronto, Ontario: University of Toronto Press.

Hanlon, B. 2017. "Suburbs." In *A Research Agenda for Cities*, edited by J. R. Short, 125–136. Cheltenham: Edward Elgar.

Hansen, H. 2017. "Assisted Technologies of Social Reproduction: Pharmaceutical Prosthesis for Gender, Race, and Class in the White Opioid 'Crisis.'" 2017. *Contemporary Drug Problems* 44: 321–338.

Jaitman, L. 2015. "Urban Infrastructure in Latin America and the Caribbean: Public Policy Priorities." *Latin American Economic Review* 24. DOI: 10.1007/s40503-015-0027-5.

Klaufus, C., P. Van Lindert, F. Van Noorloos, and G. Steel. 2017. "All-Inclusiveness Verses Exclusion: Urban Project Development in Latin America and Africa." *Sustainability* 9. DOI: 10.3390/su9112038.

Ploran, E. J., M. A. Trasciatti, and E. C. Farmer. 2018. "Efficacy and Authority of the Message Sender During Emergency Evacuations: A Mixed Methods Study." *Journal of Applied Communication Research* 46: 291–322. DOI: 10.1080/00909882.2018.1464659.

Randhawa, A., and A. Kumar. 2017. "Exploring Sustainability of Smart Development Initiatives in India." *International Journal of Sustainable Built Environment* 6: 701–710.

Lowenhaupt, R. 2016. "Immigrant Acculturation in Suburban Schools Serving the New Latino Diaspora." *Peabody Journal of Education* 91: 348–365.

Mohareb, E. A., M. C. Heller, and P. M. Guthrie. 2018. "Cities' Role in Mitigating United States Food System Greenhouse Gas Emissions." *Environmental Science and Technology* 52: 5545–5554.

Sanderson, E. W., W. D. Solecki, J. R. Waldman, and A. S. Parris, eds. 2016. *Prospects for Resilience: Insights from New York City's Jamaica Bay.* Washington, D.C.: Island Press.

Timberlake, J. M., and E. Johns-Wolfe. 2016. "Neighborhood Ethnoracial Composition and Gentrification in Chicago and New York, 1980–2010." *Urban Affairs Review* 53: 236–272.

van Beynen, P. E., M. A. Niedzielski, E. Bialkowska-Jelinska, K. Alsharif, J. and Matsusick. 2012. "Comparative Study of Specific Groundwater Vulnerability of a Karst Aquifer in Central Florida." *Applied Geography* 32: 868–877.

Watts, M. 2017. "Cities Spearhead Climate Action." *Nature Climate Change* 7: 537–538.

CONTRIBUTORS

Troy D. Abel's research and teaching interests focus on the dynamic tensions of environmental science and democratic politics in the fields of environmental justice, information disclosure, and climate governance. He is professor of environmental policy at Western Washington University's Huxley College of the Environment.

Simon A. Andrew is professor in the Department of Public Administration, University of North Texas, in Denton, Texas. He is one of the core faculty teaching in the field of public management. He has a doctoral degree from the Askew School of Public Administration and Policy, Florida State University, and a master's degree in development economics from the School of International Development, University of East Anglia (United Kingdom). He studies interorganizational collaboration in the context of disaster planning and management in theory and field settings using quantitative and qualitative empirical methods.

Viney P. Aneja is professor and codirector of graduate programs in the Department of Marine, Earth, and Atmospheric Sciences, North Carolina State University (http://go.ncsu.edu/airquality). In 2015 he was awarded the Rossby Visiting Fellow at the University of Stockholm, Sweden, and that of the Indira Foundation Distinguished Visiting Fellow at TERI University, New Delhi, India. He was recently appointed to the North Carolina Science Advisory Board. He serves on the U.S. Environmental Protection Agency's Board of Scientific Counselors (BOSC) Executive Committee (2014 to present) and served as a chair of the BOSC Subcommittee for Air, Climate, and Energy (ACE) research program (2014–2017). He is the recipient of the 2007 North Carolina Award in Science. In 1998, the Air and Waste Management Association awarded him its Frank A. Chambers Award, the association's highest scientific honor; in 1999 he became a fellow of the association; and in 2001 he received the association's Lyman A. Ripperton Award for distinguished achievement as an educator.

Miles Ballogg serves as the brownfields practice leader and economic development director for Cardno, an international consultancy for global infrastructure and social and environmental development. He assists public and private sector clients with land redevelopment programming, assessment, and remediation, specifically in environmentally blighted areas. Miles has been involved in the implementation of both federal and state brownfields programs for over 25 years, including during their inception. He is a strong advocate for environmental justice and has been recognized by the U.S. EPA and other organizations as the national leader of the "Healthfields Movement" by utilizing brownfields and other leveraged resources to improve access to healthcare for underserved communities. He successfully assisted clients in obtaining over $150 million in EPA brownfields and other redevelopment-related grants and incentives and has assisted more than 50 clients (including government agencies, nonprofits, and private sector developers) with brownfields consulting, grant writing services, and programmatic support to facilitate the transformation of environmentally impaired properties into benefits to their communities.

William H. Battye has worked for 40 years as a consultant in the environmental industry, with a specialty in air pollution and air pollution control. His work has included measurement and modeling of air pollution and technical and economic analysis of pollution controls. For 25 years, he served as managing partner of EC/R Incorporated, a firm which provided contract consulting services to the U.S. Environmental Protection Agency and other governmental bodies. He currently serves as senior scientist and engineer at SC&A Incorporated and adjunct professor in the Marine, Earth and Atmospheric Sciences department at North Carolina State University. He is a registered professional engineer and holds BS and MS degrees in chemical engineering from the Massachusetts Institute of Technology, and a PhD in air quality from North Carolina State University, Raleigh, North Carolina.

Casey D. Bray was an intern with the North Carolina Division of Air Quality and the Forsyth County Office of Environmental Assistance and Protection for four summers, where she analyzed ozone and particulate matter ambient concentrations and completed the air quality forecast (both ozone and fine particulate matter) for the state of North Carolina. She holds a BS in meteorology and a PhD in atmospheric science from North Carolina State University. She is the recipient of the 2018 John S. Irwin Award for Scientific Excellence in Atmospheric Sciences.

Robert Brinkmann is professor of geology, environment, and sustainability. He has a PhD in geography and a master's degree and bachelor's degree in geology. He has served on the writing team of the Long Island Regional Economic Development Council. His research focuses largely on human alteration of the planet. Specifically, he is interested in soil and water pollution and alteration of karst terrain. He has published dozens of books and articles on sustainability topics.

Most recently, he is the author of *Introduction to Sustainability*, one of the first major sustainability textbooks published. He is also the book series editor for environmental sustainability with Palgrave. His writing has been featured on CNN, Newsday, HuffPost, and his blog, *On the Brink*. He has appeared on a number of television (CNN, CBS News) and radio stations to talk about sustainability issues. Brinkmann was recently awarded the Eddie Mitchell Memorial Award for his work on human rights in Florida.

Vaswati Chatterjee is assistant professor in the Department of Public Administration at Villanova University in Pennsylvania. She has a doctoral degree in public administration and management from the University of North Texas, a master's in city planning from Indian Institute of Technology, Kharagpur (India), and a bachelor's in architecture from Jadavpur University (India). Her research interests include hazard mitigation and planning, involvement of local government in climate change policies, and disaster preparedness in minority communities.

Stacy Clauson is an urban planner with over 15 years' experience working for cities, counties, and regional planning organizations. Her interests are focused on building healthy, livable, and inclusive places. Recent projects examine the equity outcomes of smart growth planning in the greater Seattle, Washington.

Craig E. Colten is Carl O. Sauer Professor of Geography at Louisiana State University and has spent over 30 years working on topics in environmental geography. His early career in Illinois and Washington, D.C., centered on issues related to groundwater contamination by hazardous wastes. His recent work in Louisiana has dealt with urban and coastal hazards and community resilience. His books include the award-winning *An Unnatural Metropolis: Wresting New Orleans from Nature*, *Perilous Place and Powerful Storms: Hurricane Protection in Coastal Louisiana*, and *Southern Waters: The Limits to Abundance*.

Sarah Combs is chief executive officer of University Area Community Development Corporation Inc., a nonprofit 501(c)(3) organization that seeks to improve the economic, educational, and social levels of the more than 25,000 residents of the socially and economically vulnerable University Area community surrounding the University of South Florida. With more than 16 years of experience in the nonprofit sector, Sarah's background spans nearly every facet of the organization, from program development to land banking and real estate. Sarah was recently recognized by the *Tampa Bay Business Journal*'s "40 Under 40" and "Power 100 of Tampa Bay," WFTS-TV ABC News: Positively Tampa Bay "Game Changer," and the *Tampa Bay Times*'s "Top Business People." She is an active member and volunteer in the Tampa Bay community, serving on numerous boards, including

Safe & Sound Hillsborough, Tampa Innovation Partnership, Commission on the Status of Women Advisory Board, the Florida Alliance of CDCs, and the Advent Health Tampa Community Health Needs Assessment Committee. She is also a founding member of the Tampa Bay Community Reinvestment Association.

Yonn Dierwechter is professor of urban studies at the University of Washington, Tacoma. His books include *Smart Transitions in City-Regionalism: Territory, Politics, and the Quest for Competitiveness and Sustainability* (with Tassilo Herrschel); *Urban Sustainability Through Smart Growth: Planning, Intercurrence and the Geographies of Regional Development Across Greater Seattle*; and *Urban Growth Management and Its Discontents: Promises, Practices, and Geopolitics in US City-Regions*. Before his academic career, Dr. Dierwechter worked as a community planner in South Africa and served as a state-level legislative policy analyst. He holds a master's degree from Cornell University and a PhD from the London School of Economics.

Richard C. Feiock is internationally recognized for his expertise in local government, sustainability, and local democratic institutions, and he is a National Academy of Public Administration fellow. He holds the Jerry Collins Eminent Scholar Endowed Chair and is the Augustus B. Turnbull Professor of Public Administration and Policy in the Askew School at Florida State University. Dr. Feiock is the founding director of the FSU Local Governance Research Laboratory.

Michael H. Finewood is assistant professor in the Department of Environmental Studies and Science at Pace University in Pleasantville, New York. He is a human geographer and political ecologist who studies environmental governance, water, and urban sustainability, with explicit attention to critical geographies and justice. His current research engages watershed nonprofits to explore the challenges of water governance across politically and ecologically fragmented landscapes.

Sandra J. Garren is director of sustainability research at the National Center for Suburban Studies at Hofstra University. She is also assistant professor in the Department of Geology, Environment, and Sustainability at Hofstra University. She completed her doctorate in geography and environmental science and policy program in the Department of Geosciences at the University of South Florida, Tampa, in May 2014. She holds a bachelor's degree in earth science/geology and a master's degree in teaching. In total, Sandra has more than 25 years of experience in both academic and environmental sustainability fields which includes principal investigator, project management, and technical expertise for scientific investigations related to sustainability, greenhouse gas accounting, climate change policy, water management, energy, and environmental regulation. Her research is focused on sustainability, water management, and climate policy issues both globally and nationally (including federal, state, and local governments) and is interdisciplinary and specifically focused on applied science and policy that solve problems and find solutions to the negative impacts of climate change and other environmental challenges. She currently teaches courses in sustainable development, sustainability theory, sustainable energy, and geospatial applications in sustainability.

Melissa M. Grigione is professor in the Biology Department at Pace University in New York. She earned her PhD in ecology at UC-Davis and concentrates her research in mammalian spatial ecology, emphasizing conservation biology for species whose populations have been seriously altered as a consequence of habitat degradation and fragmentation. She is coauthor of numerous peer-reviewed papers and articles and the recipient of more than a dozen grants and fellowships.

John Harner is professor at the Department of Geography and Environmental Studies University of Colorado in Colorado Springs, Colorado. John is an urban and historical geographer with a research focus on the processes of place creation in Colorado Springs and Guadalajara, Mexico. He recently collaborated on a digital humanities project that brings Colorado Springs' historical landscapes to the public at the Colorado Springs Pioneers Museum (see http://www.cspm-storyofus.com), much of which will be incorporated into his forthcoming book on the city.

Mathew K. Huxel received his MBA in entrepreneurship from Southern New Hampshire University in 2014. While pursuing a second master's degree, Mathew was a sustainability associate for the Town of Hempstead's Department of Conservation and Waterways working on issues related to energy benchmarking, while also working as a graduate research assistant at Hofstra's National Center for Suburban Studies. In 2019, he received his second graduate degree, a master of art in sustainability at Hofstra University, where he focused on the adoption and implementation of sustainability policy at the city level.

Mike Johnson is retired from the University of California, Davis, where his research focused on population ecology and evolutionary biology. He received his PhD from the University of Kansas, where he worked on the evolution of dispersal in mammals. He is the author of numerous peer-reviewed publications and book chapters.

Gabrielle R. Lehigh is a PhD student in applied anthropology at the University of South Florida, where she earned her master's degree in the same field studying an EPA-funded brownfields redevelopment project. Her research focuses on environmental justice issues, especially community engagement in the redevelopment of brownfields as well as natural resource extraction and energy production. Since 2016, she has served as a research associate at the Center for Brownfields Research and Redevelopment at the University of South Florida. She has presented her research at a variety of local, regional, and national conferences including those of the American Anthropological Association, the Society for Applied Anthropology, the Florida Brownfields Association, the Florida Anthropological Society, and the Interdisciplinary Symposium for Qualitative Methodologies. She has recently been featured on a series of radio stations and podcasts to talk about community development and environmental remediation.

Elizabeth Mattiuzzi is senior researcher in community development at the Federal Reserve Bank of San Francisco. Her research focuses on regional transportation and housing governance, equity, and economic opportunity. She holds a PhD in city and regional planning from the University of California, Berkeley.

Sean McGreevey is dean of students at Bellarmine University in Louisville, Kentucky. His scholarly interests include leadership development and experiential learning. He is an American Canoe Association–certified whitewater kayak instructor and a graduate of the National Outdoor Leadership School.

Susan M. Opp is professor of public policy and administration at Colorado State University. She has a PhD in urban and public affairs, a master's degree in public administration, and a bachelor's degree in economics. She has served in several important public service roles related to sustainability and economic development including serving as Pracademic Fellow at the Environmental Protection Agency's Office of Policy, a panel manager for the Community Economic Development program in the Department of Health and Human Services, and a research associate for the Environmental Finance Center at the University of Louisville. Her research focuses largely on local sustainability considerations in the United States with a particular focus on the economic aspects of sustainability. She is the author of a number of books, articles, and professional reports on sustainability, local economic development, and environmental policy. Most recently, she is the coauthor of *Performance Measurement in Local Sustainability Policy*, the coeditor of a two-volume series, *Environmental Issues Today*, and the coauthor of *Local Economic Development and the Environment: Finding Common Ground*.

Michaela C. Peterson is a PhD student in earth and environmental science at Vanderbilt University. She earned her BA in biology from Skidmore College, and her MS in environmental science at Pace University. Her current research is focused on the foraging energetics and thermoregulatory behavior of neotropical ungulates.

Benjamin L. Ruddell is associate professor and director of the School of Informatics, Computing, and Cyber Systems at Northern Arizona University, the president of Ruddell Environmental consulting, and the director of the FEWSION project. His PhD is in civil and environmental engineering from the University of Illinois at Urbana-Champaign. His professional experiences are in the fields of civil engineering, water resources, systems analysis, ecology/ecohydrology, and engineering research and education in an interdisciplinary university setting. He works with a variety of federal, local, and private partners to accomplish cutting-edge projects. His research interests fall broadly in the area of the quantification and management of complex coupled natural-human systems, including regional water and climate systems strongly influenced by the human economy and society—such as in cities, energy, and agriculture. His professional goals are the advancement of the science and management of complex systems and excellence in education in a university setting.

Richard R. Rushforth is assistant research professor in the School of Informatics, Computing, and Cyber Systems at Northern Arizona University. His research focuses on big data modeling of food, energy, and water systems to further the understanding of complex, coupled natural-human systems. His PhD is in civil, environmental, and sustainable engineering from Arizona State University. He also holds degrees from the University of Oxford (MSc, water science, policy, and management) and the University of Arizona (MS, soil, water, and environmental science; BS, environmental science) as well as an MBA from the W. P. Carey School of Business at Arizona State Universi-ty.

Debra Salazar is professor of political science and affiliate professor of environmental studies at Western Washington University. Her research and teaching center on the relation between justice and environment. Recent projects examine state policies related to inequality in the distribution of industrial air toxins and exposure to agricultural pesticides.

Ronald Sarno is professor in the Biology Department at Hofstra University in New York. He earned his PhD in ecology/evolutionary biology at Iowa State University. Much of his past research has focused on the population ecology, evolution, and behavior of mammals, especially the wild South American camelids due to his interest in mating systems, social behavior, population dynamics, and conservation.

Mallory Thomas is a geographic information systems specialist working with the Marine Corps. She earned a Master of Science degree from the Department of Geography and Anthropology in the College of Humanities and Social Sciences with a graduate minor from the Department of Environmental Sciences in the

College of Coast and Environment at Louisiana State University in Baton Rouge, Louisiana. Her graduate research focused on data science, spatial modelling, public health, urban air quality, and land cover and land use.

Carolina A. Urrea is a graduate student in the master of sustainability program at Hofstra University in New York and earned a bachelor of science degree in biology at Molloy College. Carolina is a graduate student researcher at Hofstra University's National Center for Suburban Studies. Her graduate research is focused on green space within suburban communities on Long Island, New York.

Pornpan Uttamang worked as a scientist with the Industrial Estate Authority of Thailand (IEAT) where she specialized in air pollution and air pollution control for industrial estates. She holds a BS in science and an MS in environmental science from Chulalongkorn University in Thailand. She was a participant in the International Visitor Leadership Program (IVLP) in atmospheric science, supported by the Bureau of Education and Cultural Affairs of the United States Department of State, U.S. embassy. She is currently a PhD candidate in atmospheric science at North Carolina State University, funded by the Royal Thai Government. She is the recipient of the 2019 Ralph I. Larsen Air Quality Research Award.

E. Christian Wells is professor of anthropology and director of the Center for Brownfields Research and Redevelopment at the University of South Florida, where he has served previously as founding director of the Office of Sustainability and deputy director of the School of Global Sustainability. He is an environmental anthropologist with a diverse portfolio of research projects aimed at improving human and environmental health outcomes of re/development efforts in marginalized communities. He has undertaken social and environmental science research throughout the United States, Central America, and the Caribbean with over $7 million in funding from the National Science Foundation, the National Institute of Justice, the U.S. Environmental Protection Agency, and other organizations. He has written or edited 10 books and journal issues and more than 100 scientific articles and essays. His work has been featured in the *New York Times* and *New Scientist*, among other media. He has received numerous awards for research, teaching, and community-engaged service, including the Black Bear Award from the Sierra Club of Tampa Bay "in recognition of outstanding dedication to sustainability and the environment."

INDEX

Printed in the United States
By Bookmasters